Technology and Anarchy

Postphenomenology and the Philosophy of Technology

Series Editors: Robert Rosenberger, Peter-Paul Verbeek, Don Ihde

As technologies continue to advance, they correspondingly continue to make fundamental changes to our lives. Technological changes have effects on everything from our understandings of ethics, politics, and communication, to gender, science, and selfhood. Philosophical reflection on technology can help draw out and analyze the nature of these changes, and help us to understand both the broad patterns of technological effects and the concrete details. The purpose of this series is to provide a publication outlet for the field of philosophy of technology in general, and the school of thought called "postphenomenology" in particular. The field of philosophy of technology applies insights from the history of philosophy to current issues in technology and reflects on how technological developments change our understanding of philosophical issues. Postphenomenology is the name of an emerging research perspective used by a growing international and interdisciplinary group of scholars. This perspective utilizes insights from the philosophical tradition of phenomenology to analyze human relationships with technologies, and it also integrates philosophical commitments of the American pragmatist tradition of thought.

Recent Titles in This Series

Technology and Anarchy: A Reading of Our Era, by Simona Chiodo

Feedback Loops: Pragmatism about Science & Technology, edited by Andrew Wells Garnar and Ashley Shew

Sustainability in the Anthropocene: Philosophical Essays on Renewable Technologies, edited by Róisín Lally

Unframing Martin Heidegger's Understanding of Technology: On the Essential Connection between Technology, Art, and History, by Søren Riis, translated by Rebecca Walsh

Postphenomenological Methodologies: New Ways in Mediating Techno-Human Relationships, edited by Jesper Aagaard, Jan Kyrre Berg Friis, Jessica Sorenson, Oliver Tafdrup, and Cathrine Hasse

Animal Constructions and Technological Knowledge, by Ashley Shew

Using Knowledge: On the Rationality of Science, Technology, and Medicine, by Ingemar Nordin

Postphenomenology and Media: Essays on Human–Media–World Relations, edited by Yoni Van Den Eede, Stacey O'Neal Irwin, and Galit Wellner

Technology and Anarchy

A Reading of Our Era

Simona Chiodo

LEXINGTON BOOKS
Lanham • Boulder • New York • London

Published by Lexington Books
An imprint of The Rowman & Littlefield Publishing Group, Inc.
4501 Forbes Boulevard, Suite 200, Lanham, Maryland 20706
www.rowman.com

6 Tinworth Street, London SE11 5AL, United Kingdom

British Library Cataloguing in Publication Information Available

Library of Congress Control Number: 2020945945

ISBN 978-1-7936-3294-4 (cloth)
ISBN 978-1-7936-3295-1 (electronic)

Contents

Introduction vii

1 The Evolution of Prometheus, from Past to Present to Future 1

2 *Homo anarchicus* 27

3 Reading Our Technological Era 51

4 Radical Externalization 73

5 Farewell to Humans? 97

6 The Most Radical Form of Anarchism We Have Ever Experienced 109

Conclusion 129

References 133

Index 145

About the Author 151

Introduction

The aim of this book is to offer a novel reading of our technological era as the last step of a precise trajectory of the Western culture. More precisely, I believe that the technological turn we are facing may be promisingly read as the most radical form of anarchism we have ever experienced. We are not only removing the role of the expert as a mediator but also trying, for the first time in our history, to replace the role of a transcendent god itself by creating ourselves, especially through computer engineering, a totally immanent technological entity characterized by the typical ontological prerogatives of the divine: omnipresence (by being everywhere), omniscience (by knowing everything, especially about us), and omnipotence (by having power, especially over us), as well as inscrutability.

I am a philosopher, and, therefore, I shall use the tools of my discipline, from the history of philosophy to its theoretical tools. But I shall also try to use another kind of tool from my background: the arts and, in particular, literature, from ancient myths to contemporary novels, since I believe that they may be invaluable in an attempt to understand exceedingly complex cultural phenomena both synchronically and diachronically, especially when the focus is on what is ongoing. In particular, I hope to be able to use both philosophy and literature to position our technological era fairly precisely on a historical trajectory rooted in the very beginning of our culture as Western humans and, therefore, to add useful tools in an attempt to understand our technological era as best we can.

Trying to achieve this aim in less than two hundred pages necessarily means selecting one of possible alternative views. As such, what I shall try to do is to be as intellectually honest as possible, that is, I shall select a view from its possible alternatives not because I think that there are no other promising views, but because I honestly think that the view I shall propose may be quite promising (as well as quite promisingly complementary to its possible alternatives).

In the title of this book, *Technology and Anarchy*, one can find both its subject (the notion of technology) and its nominal predicate, as it were (the notion of anarchy). The actual nominal predicate I shall use in this book to qualify our technological era is the notion of anarchism rather than the notion of anarchy, the former being the radicalization of the latter. However, both the publisher and I believe that, given the historical trajectory I shall analyze in this book, it is useful to start from the meaning of anarchy as what actually underpins, both chronologically and logically, the meaning of anarchism as its radicalization (see especially the second and sixth chapters).

Even though this book ends with an extensive list of references, it cannot be taken to be exhaustive, given the complexity of the cultural trajectory they focus on. In any case, I hope that the references can be useful as, at the very least, a jumping off point for further investigations that may develop particular details of the general trajectory I want to try to describe in the following pages.

As for the major steps of the trajectory I shall work on, I shall start from the analysis of the historical evolution of the figure of Prometheus as a powerful symbol of the meaning of technology in the Western culture, from its genesis to our technological era (first chapter: *The evolution of Prometheus, from past to present to future).* I shall try to argue that technology is something that has constituted human nature from the very beginning, and that the kind of challenge that humans undertake through technology is changing and, specifically, improving, their ontology, by moving from a kind of earthly dimension to a kind of divine dimension. More precisely, I shall try to argue that the most essential meaning of technology has a surprising correlation with the relationship between the human and the divine. The modern evolution of the figure of Prometheus may be read as a representation of the phase in which technology allows humans to challenge the divine by creating living human matter from nonliving matter. Moreover, a possible contemporary evolution of the figure of Prometheus may be read as a representation of the phase in which technology allows humans to challenge the divine by creating the divine itself as a technological entity characterized by the typically divine prerogatives of omnipresence, omniscience, and omnipotence. More precisely, I shall try to argue that we are creating a divine that is totally immanent, which is the essential reason why our technological era may be read as the most radical form of anarchism we have ever experienced. Whereas the human even exceeds the divine by technologically creating it, we are facing another kind of ontological move: the more technology develops, the more indiscernible the ontological difference between the human and the technological becomes.

As I shall try to argue, the three major steps of the historical evolution of the figure of Prometheus correspond, most meaningfully, to the three major steps of the historical evolution of the foundation of the Western culture from an epistemological perspective: the invention of ideality and the evolution of its relationship with reality (second chapter: *Homo anarchicus*). I think that reflecting upon this epistemological evolution is crucial to try to understand the trajectory of technology and, as a result, our technological era. From an epistemological perspective, the Western culture is founded on the invention of ideality as the perfect counterpart of reality, that is, as its privileged model and, therefore, as the privileged working method of technology: if we want to make an artifact that is as good as possible and increasingly improves, then we must refer to its ideal model. However, the special form of epistemological anarchism that characterizes our technological era means not only that the superior ontology of ideality experiences a severe crisis but also that we are replacing it by technologically creating an ontology that is, outwardly, superior by being omnipresent, omniscient, and omnipotent but, inwardly, inferior by being totally immanent and, in particular, a totally immanent human creation. This special form of epistemological anarchism is the most radical form of anarchism we have ever experienced precisely because what we are actually doing is creating reality as a unique ontological dimension by creating a special kind of divine that embodies, outwardly, the typical ontological qualities of ideality, starting from its perfection and, inwardly, the typical ontological qualities of reality, from its imperfection to its solipsism, contingency, accidentality, idiosyncrasy, and randomness.

After having analyzed the founding epistemology of our trajectory, I shall take into account concrete examples that characterize our technological era, from the cases of algorithm and health to the cases of the web, money, and politics (third chapter: *Reading our technological era*). I shall try to argue that their philosophical meanings are frequently related to a severe crisis of the notion of individual responsibility. More precisely, I shall try to argue that we are trying to use technology to free ourselves from individual responsibility as one of the most typically human prerogatives, at least in the Western culture from Aeschylus to Shakespeare to Pirandello. Moreover, we are trying to use technology as a special kind of bureaucracy that coincides, again, with a technological entity as an immanent divine obtaining another typically divine prerogative, that is, inscrutability. Thus, radical anarchism seems to be our way of removing what has been thought of as our best, as well as most challenging, guide: an individual responsibility as an ideal rule to measure up to. The concrete examples taken into account shall show this phenomenon in terms of the use of technology to remove the role of both the expert guiding us and the ideal rule to measure up to.

I shall develop this argument by stressing that we seem to revolutionize our identity as Western humans, in that we seem to choose to become radically anarchistic by externalizing epistemological and ethical decisions, as well as the burden of being responsible for them, by creating a totally immanent technological divine that makes epistemological and ethical decisions for us (fourth chapter: *Radical externalization*). For millennia, we have been thinking of ourselves as bearers of the burden of responsibility for our actions as the result of our decisions even though they may have tragic consequences on both ourselves and others. Yet, our technological era may be read precisely as our greatest struggle to try to shift this burden from ourselves to technology by increasingly externalizing everything from any epistemological decision to any ethical decision—more precisely, we are increasingly shifting our decisions from our minds, as well as from our bodies, to the algorithmic machines we create. In particular, I shall distinguish two notions: externalizing content, which may be defined as an externalization, and externalizing abilities, which may be defined as a meta-externalization. And I shall try to argue that our technological era coincides especially with a radical meta-externalization. However, if it is true, as I shall try to argue, that meta-externalizing (abilities) is more crucial than externalizing (content) when it comes to deciding on the nature of the identity of a human being, then it is also true that the most crucial reason why a radical meta-externalization of our abilities is likely to be a critical choice is not only a matter of neutralizing our abilities but also, and above all, a matter of neutralizing the way we understand, decide, and plan our identities as human beings. Indeed, we are massively meta-externalizing the very core of living human lives as they have been lived for millennia, that is, as a matter of being courageous bearers of the burden of responsibility for our decisions. Finally, through concrete examples, I shall try to show that, most interestingly, meta-externalizing anything from the human to the technological is likely to cause the de-evolution of the former, paradoxically enough, into a pretechnological ontological status.

From an epistemological perspective, I believe that the phenomena of radical externalization and meta-externalization may also be promisingly analyzed in terms of a progressive hypertrophization of the notion of *logos* that, not coincidentally, seems to experience its most radical form in our technological era (again, fourth chapter). The essential tool through which we have been bearing the burden of individual responsibility has been precisely the notion of *logos*, which has been equally essential for defining the very core of Western human identity for millennia. Indeed, *logos* is what makes Western humans understand what is and predict what will be and, therefore, help decide and act rationally. As I shall try to argue, the history of the Western culture may be read as follows:

1. We have been increasingly entrusting our control of reality, that is, our understanding of what is and our predictions of what will be, to *logos*, which is a strict form of rationality.
2. We have been increasingly transforming *logos* into an even stricter form of rationality, which is logic—we have been increasingly hypertrophying *logos*.
3. More recently, we have been starting to think, first, that the best form of *logos* is its strictest form and, second, that its strictest form is computation.
4. Finally, we have been starting to believe that the best form of computation is its radical meta-externalization from our minds (considered less powerful) to our machines (considered more powerful) and, in particular, to our algorithmic machines (considered the most powerful), and, therefore, that we should both radically externalize and radically meta-externalize.

However, our minds' abilities exceed rationality, which exceeds *logos*, which exceeds computability. Thus, are we actually sure that algorithmic abstraction is always more powerful than mental abstraction? Do we sufficiently ask ourselves how we can improve our minds' powers? And do we financially invest in this idea to a sufficient extent? It is fairly clear that both our millennial history of progressive hypertrophization of *logos* and our more recent history of its progressive meta-externalization do not help. Indeed, their union even seems to establish a kind of dogma, according to which investigating outside of *logos* is not only unintelligent from an epistemological perspective but also unworthy from an ethical perspective. Thus, the logical paradox is the following: first, we have been founding the very core of our identity on the exercise of *logos* itself; second, we have been increasingly radicalizing the relationship between the former and the latter; third, we have been increasingly meta-externalizing the latter, especially in our technological era; fourth, what will constitute the very core of our identity, especially in the future?

My next step shall be focusing on reading our technological era as the most radical form of anarchism we have ever experienced precisely because anarchism may be our best way of trying to give away our most unbearable burdens (fifth chapter: *Farewell to humans?*). In particular, I shall try to argue that one of the most important reasons of our creation of a technological bureaucracy that becomes increasingly opaque, rigid, and hard to negotiate with is that we are creating a kind of technological scapegoat to replace us even more than a kind of technological servant to replace us (sixth chapter: *The most radical form of anarchism we have ever experienced*). Yet, externalizing faults by creating a technological scapegoat means externalizing autonomy, which is another essential notion for the Western modern and contemporary culture, as Immanuel Kant masterfully teaches. By referring to Kant's philosophy, I shall ask the following: can we define our technological era as the technological creation of a god that is actually the greatest *automaton materiale*, which, by being always with us through its immanent omnipresence, omniscience, and omnipotence, as well as through its bureaucratic inscrutability, can always be our technological scapegoat and servant, always freeing us from the most unbearable burdens of autonomy? Answering this question shall mean arguing the following:

1. Since the twentieth century, we have started to think of ourselves as not only equal to the divine by obtaining an analogous power of creation but also a replacement for the divine by obtaining its very power of creation, that is, the power of creating the world itself, meaning the world as it is for us.
2. In our technological era, we are going a step further, in that we are replacing the divine by obtaining the power of technologically creating the divine itself as a totally immanent human creation that can be our scapegoat and servant, always freeing us from the most unbearable burdens of autonomy.

The European human being (and, by extension, the Western human being) has been thought of for millennia as an autonomous individual bearing the burdens of being rational and ethical—we have been thinking of ourselves for millennia as autonomous individuals whose values are measured in terms of our abilities to understand, decide, and act, first, rationally and, second, ethically. This is the very core of our identity as European humans (and, by extension, Western humans)—and this is precisely what we are radically externalizing and meta-externalizing in our technological era.

Finally, I shall reflect upon the possibility of also thinking of our technological era as an attempt to obtain the typically divine prerogative of existing in a kind of privileged everlasting imperturbability. If this makes sense, then it may be possible to add that we seem to increasingly meta-externalize from ourselves to technology even the ability to evolve, which means that we seem to replace our changing and evolving with technology's changing and evolving. Indeed, the kind of ontological status we seem to desire for technology is complex. On the one hand, it is becoming divine by obtaining the most typically divine prerogatives. On the other hand, it is becoming human by obtaining the human abilities *par excellence*. And, as technology obtains both divine and human prerogatives, we become more and more radically anarchistic. Indeed, even from an etymological perspective, as I shall show in the sixth chapter, there is a surprising coincidence between the words defining "*automaton*" and technological "automation," on the one hand, and the words defining anarchism, on the other hand: radically replacing autonomy with technological "automation" means becoming radically anarchistic, since the meaning of autonomy is replaced by the randomness meant precisely by the etymology of the technological "automation" that may become too omnipresent, omniscient, omnipotent, and inscrutable to save our autonomy.

As I write this book, we are facing the coronavirus emergency, which can act as a kind of accelerator for the trajectory I shall try to describe in this book (*Conclusion*). Indeed, it seems to accelerate anarchism, both in terms of the crisis of the expert guiding us (whether the expert is a scientist or a politician) and in terms of the use of technology (from the fake news haunting every form of social media to our use of technology as a kind of do-it-yourself *passe-partout* replacing every form of guide). And, while we are falling into the most radical form of anarchism we have ever experienced, nature is showing us to be the most powerful, in any case, by being a virus that can perfectly use our most sophisticated technology as its most powerful tool to quickly spread to all corners of the planet and get to each and every one of us.

Chapter 1

The Evolution of Prometheus, from Past to Present to Future

THE ANCIENT PROMETHEUS

Why should we start a book about our technological era by talking about Prometheus? The answer I can offer is the following: because, if the issue at stake is exceedingly complex, then we need a diachronic perspective in order to understand it. Our technological era is exceedingly complex not only because it is quickly changing the very notion of human being but also because it is ongoing. And trying to understand the meaning of something exceedingly complex from a totally internal perspective is always destined to be unsuccessful, as trying to understand a disease by looking at the symptom without looking at its cause. In our case, we shall use the historical evolution of the figure of Prometheus as a powerful symbol to allow us to give our technological era a precise position on a map, as it were, and, therefore, to try to understand it as best we can, that is, not by identifying a symptom but by identifying a cause and, consequently, the disease itself.

In the Western culture, the figure of Prometheus has been represented for millennia as the most powerful symbol of the relationship between humans and technology, from mythology to art to philosophy[1]. By taking into account the major sources, both synchronically and diachronically, I shall try to focus not only on the meaning of the figure of Prometheus, as well as on its historical evolution, but also on its trajectory, which, as we shall see, seems to be fairly precise and most illuminating. More precisely, I shall take into account three major steps of the historical evolution of the figure of Prometheus: the ancient Prometheus, as he is described especially by Hesiod, Aeschylus, Plato, and Ovid; the modern Prometheus, as he is described especially by Mary Shelley (but also through a look at Percy Bysshe Shelley); and, finally, a possible contemporary Prometheus, as he may be inferred especially from contemporary art. Of course, this is a selection of the most meaningful versions of the figure of Prometheus since my aim is not its philological reconstruction but its possible philosophical use.

Before taking into account the historical evolution of the figure of Prometheus, let us consider the etymological meaning of the name "Prometheus" itself, which derives from the combination of the Greek words πρό (before) and μανθάνω (knowing) and, therefore, means "predicting," which is actually one of Prometheus' prerogatives. Also Hesiod, in his *Theogony*, helps us understand the etymological meaning of the name "Prometheus" by referring to the opposing natures of Prometheus and his brother Epimetheus: the latter is said to be "of-erring-mind" (Hes. *Theog.* 510), whereas the former is said to be "artful [. . .], full of various wiles" (Hes. *Theog.* 510), "wily-minded" (Hes. *Theog.* 520), "helpful" (Hes. *Theog.* 610), and "very knowing" (Hes. *Theog.* 615).

By highlighting their different mental attitudes, Hesiod confirms Prometheus' etymological meaning versus Epimetheus' etymological meaning, which is the opposite of the former in that it derives from the combination of the Greek words ἐπί ("after") and μανθάνω ("knowing"). More precisely, the reason why Prometheus has, in general, powerful mental attitudes and, in particular, the ability to predict, also in terms of being far-sighted, is partly clarified by a myth that is not told in Hesiod's *Theogony* but in Plato's *Protagoras* (see Plat. *Prot.* 320 d ff.): when Epimetheus is asked by the Olympians to assign a given number of good qualities to the living creatures, he gives animals all the good qualities at his disposal totally neglecting humans, whereas Prometheus fixes his mistake by giving humans Hephaestus' fire and Athena's wisdom in the arts, including mastery over fire itself. Thus, Prometheus' ability to predict has to do with something fairly precise: we may say that he is able to predict that, if humans have no good quality, being "naked, unshod, unbedded, unarmed" (Plat. *Prot.* 321 c) and deprived of the animals' strength and speed, then what they will need in order to survive the animals themselves is something divine and, in particular, the ability to master fire. This is a point greatly important: the reflection upon the etymology of the name "Prometheus" leads us to think that what distinguishes humans from the other living creatures is that, whereas the latter have in themselves, that is, in their very bodies, what they need in order to survive, the former need something else in order to survive, that is, the intelligent mastery of fire. Yet, as we shall see, this does not mean that the intelligent mastery of fire, that is, technology, is something extraneous to human nature—on the contrary, this means that technology is a constituent part of the human nature at its very core.

What everyone knows about the myth of Prometheus is that he steals the divine fire from the gods to give it to humans in order to help them survive.

But the myth is more complex, especially when it is told by philosophers. Plato specifies that there are two major consequences of obtaining the divine fire:

1. The first is described as follows: "And now that man was partaker of a divine portion, he, in the first place, by his nearness of kin to deity, was the only creature that worshipped gods, and set himself to establish altars and holy images" (Plat. *Prot.* 322 a).
2. And the second is described as follows: "secondly, he soon was enabled by his skill to articulate speech and words, and to invent dwellings, clothes, sandals, beds, and the foods that are of the earth" (Plat. *Prot.* 322 a).

Plato is arguing something essential, for at least two reasons. First, the meaning of the intelligent mastery of fire concerns humans' ability to survive both the animals and the weather conditions and even to rule nature (both by "articulat[ing] speech and words" and by "invent[ing] dwellings, clothes, sandals, beds, and the foods"), but it is not reducible to this ability. Indeed, the intelligent mastery of fire has a second meaning, that is, a special relationship between the human and the divine: what distinguishes humans from the other living creatures is their being "partaker[s] of a divine portion," their "nearness of kin to deity." As we shall see, this special relationship between the human and the divine has other essential consequences and, specifically, challenging ones, which are not limited to "worship[ing] gods" and "establish[ing] altars and holy images." Second, by choosing a precise logical order, Plato seems to argue that this special relationship between the human and the divine is the most essential meaning of the intelligent mastery of fire—we may start to reflect upon something we shall develop in the following pages: what Plato seems to argue is that the most essential meaning of technology has a surprising correlation with the relationship between the human and the divine.

We have reflected upon Hesiod's founding description of Prometheus by considering his mental attitudes ("artful [. . .], full of various wiles," "wily-minded," and "very knowing"). But, as we have seen, there is another attitude described by Hesiod: Prometheus is "helpful." Of course, Prometheus is "helpful" to humans, and Hesiod tells us why by adding something important to our knowledge of him:

> When the gods and mortal men were contending at Mekone, then did he [Prometheus] set before him [Zeus] a huge ox, having divided it with ready mind, studying to deceive the wisdom of Zeus. For here, on the one hand, he deposited the flesh and entrails with rich fat on the hide, having covered it with the belly of the ox; and there, on the other hand, he laid down, having

> well-disposed them with subtle craft, the white bones of the ox, covering them with white fat. Then it was that the father of gods and men addressed him, "Son of Iapetus, far-famed among all kings, how unfairly, good friend, you have divided the portions." Thus spoke rebukingly Zeus, skilled in imperishable counsels. And him in his turn wily Prometheus addressed, laughing low, but he was not forgetful of subtle craft: "Most glorious Zeus, greatest of ever-living gods, choose which of these your inclination within your breast bids you." He spoke subtlety: but Zeus knowing imperishable counsels was aware, in fact, and not ignorant of his guile; and was boding in his heart evils to mortal men, which also were about to find accomplishment. Then with both hands he lifted up the white fat. But he was incensed in mind, and wrath came around him in spirit, when he saw the white bones of the ox arranged with guileful art. (Hes. *Theog.* 535–55)

Thus, Prometheus is "helpful" to humans in a radical way. His help is not limited to doing something to provide humans with an advantage—on the contrary, Prometheus' helping humans means doing something that provides them with an advantage that, at the same time, disadvantages the gods. Interestingly enough, Zeus' first reaction to Prometheus' deceit is a punishment against humans and not against Prometheus—and the punishment against humans is exceedingly meaningful: "from that time forward, ever mindful of the fraud, he [Zeus] did not give the strength of untiring fire to wretched mortal men, who dwell upon the earth" (Hes. *Theog.* 560). Later, when Prometheus steals fire to bring it back to humans, Zeus finds out and punishes him severely, being continuously tortured by an eagle that eats his liver.

This logical order is most illuminating: first, Prometheus deceives Zeus; second, Zeus punishes humans by depriving them of fire; third, Prometheus steals fire to bring it back to humans; fourth, Zeus punishes Prometheus. What is most illuminating is that this logical order shows that Zeus' crucial issue has nothing to do with being deceived by Prometheus but, paradoxically enough, with competing with humans. Indeed, Zeus' first reaction to Prometheus' deceit is a punishment against humans—more precisely, a most meaningful punishment against humans: depriving them of fire, that is, technology. Thus, we can argue that the symbolic meaning of the myth of Prometheus correlates with the following idea: technology is precisely what gives humans the possibility of competing with the gods.

Indeed, if there is an imbalance between them and, in particular, if humans get an advantage over the gods, then the gods' first reaction is to deprive humans of technology (whereas, the punishment of the actual deceiver, that is, Prometheus, comes later). Thus, we can add something more precise to our argument that the most essential meaning of technology surprisingly correlates with the relationship between the human and the divine: its foundation is the fact that technology seems to coincide with the tool that can bridge the gap between humans and the gods—technology seems to coincide with the tool used by humans in their attempt to become divine.

It is no coincidence that the opposite imbalance, that is, the unbridgeable gap between humans and the gods before the former obtained the ability to master fire, is impressive as it is described by the myth and, in particular, by Aeschylus in his *Prometheus Bound* when Prometheus explains what human life was like before his gift:

> Listen to the miseries that beset mankind—how they were witless before and I made them have sense and endowed them with reason. I will not speak to upbraid mankind but to set forth the friendly purpose that inspired my blessing. First of all, though they had eyes to see, they saw to no avail; they had ears, but they did not understand; but, just as shapes in dreams, throughout their length of days, without purpose they wrought all things in confusion. They had neither knowledge of houses built of bricks and turned to face the sun nor yet of work in wood; but dwelt beneath the ground like swarming ants, in sunless caves. They had no sign either of winter or of flowery spring or of fruitful summer, on which they could depend but managed everything without judgment, until I taught them to discern the risings of the stars and their settings, which are difficult to distinguish. Yes, and numbers, too, chiefest of sciences, I invented for them, and the combining of letters, creative mother of the Muses' arts, with which to hold all things in memory. I, too, first brought brute beasts beneath the yoke to be subject to the collar and the pack-saddle, so that they might bear in men's stead their heaviest burdens; and to the chariot I harnessed horses and made them obedient to the rein, to be an image of wealth and luxury. It was I and no one else who invented the mariner's flaxen-winged car that roams the sea. Wretched that I am—such are the arts I devised for mankind. (Aesch. *PB* 436–70)

The consequence of Prometheus' gift, that is, technology as it is symbolized by the intelligent mastery of fire, is impressive. Humans without technology are, both literally and figuratively, "like swarming ants," "dwelt beneath the ground [. . .] in sunless caves," that is, like animals who cannot, literally, see the sunlight and, figuratively, discern (they are "witless," do "not understand," are "without purpose," "in confusion," and "without judgment"). On the contrary, humans with technology are the animals' masters since, unlike the animals, they can "reason," "discern," and "distinguish" through both the most concrete forms of technology ("the yoke," "the collar," "pack-saddle," "the chariot," "the rein," and the "flaxen-winged car") and the most abstract forms of technology ("numbers," the "sciences," "letters," and the "arts"). Moreover, Aeschylus shows that the series of consequences of the first two forms of technology is even more impressive:

> Hear the rest and you shall wonder the more at the arts and resources I devised. This first and foremost: if ever man fell ill, there was no defence—no healing food, no ointment, nor any drink – but for lack of medicine they wasted away, until I showed them how to mix soothing remedies with which they now ward off all their disorders. And I marked out many ways by which they might read the future, and among dreams I first discerned which are destined to come true; and voices baffling interpretation I explained to them, and signs from chance meetings. The flight of crook-taloned birds I distinguished clearly—which by nature are auspicious, which sinister—their various modes of life, their mutual feuds and loves, and their consortings; and the smoothness of their entrails, and what color the gall must have to please the gods, also the speckled symmetry of the liver-lobe; and the thigh-bones, wrapped in fat, and the long chine I burned and initiated mankind into an occult art. Also I cleared their vision to discern signs from flames, which were obscure before this. Enough about these arts. Now as to the benefits to men that lay concealed beneath the earth— bronze, iron, silver, and gold—who would claim to have discovered them before me? No one, I know full well, unless he likes to babble idly. Hear the sum of the whole matter in the compass of one brief word—every art possessed by man comes from Prometheus. (Aesch. *PB* 475–505)

Surprisingly enough, Aeschylus' Prometheus shows us what humans, that is, any human being from the past to our technological era, can have through technology. First, humans can cure diseases. Second, humans can predict the future. Third, humans can exploit nature. Let us analyze the essential meaning of the three major consequences of technology. What they share is precisely what makes humans closer to the gods than they are to the animals. Humans can cure diseases, which means that they can aspire not only to increasingly live better but also to increasingly live longer (and we know that this is precisely what happened and keeps happening even more ambitiously through programs that try to prevent us from aging, falling ill, and, finally, dying—for instance in the case of the SENS project[2]. See also Domingos 2015; Harari 2016). Humans can predict, which means that they can aspire to see not only the present, as the animals can, but also the future, as the gods can (and we know that this is precisely the most essential aim of science, which shares with what Aeschylus describes, from reading dreams to reading natural signs, the aim of "read[ing] the future," even though science has moved from dreams and natural signs to ideal models). And humans can both exploit and rule over nature, which means that they can aspire not only to take natural materials but also to shape materials in order to create something else (and we know that this is precisely the most essential aim of engineering, whose ability to create is exponentially increasing in our technological era. See at least Arthur 2009; Franssen, Lokhorst and van de Poel 2018;

Terrone 2019). Thus, Aeschylus' Prometheus shows us that technology is what allows humans to challenge, first, the limits of human lifespan, second, the limits of human knowledge, and, third, the limits of human creation. We may summarize by saying that what Aeschylus' Prometheus shows us is that the most essential meaning of technology is, again, an attempt to bridge the gap between humans and the gods by making the former try to obtain the latter's typical prerogatives and, specifically, being everlasting, omniscient, and creators.

It is no coincidence that Prometheus' gifts to humans are divine: Hephaestus' fire and Athena's wisdom in the arts, including mastery over fire itself. More precisely, according to Plato, "Prometheus, in his perplexity as to what preservation he could devise for man, stole from Hephaestus and Athena wisdom in the arts together with fire—since by no means without fire could it be acquired or helpfully used by any—and he handed it there and then as a gift to man" (Plat. *Prot.* 321 c–d). Thus, we may even say that there are three divine elements given to humans: first, Hephaestus' wisdom; second, Athena's wisdom; third, fire itself, which is the condition for the existence of wisdom, in that wisdom is totally neutralized without fire ("by no means without fire could it be acquired or helpfully used by any"). Plato's words are crucial, since they introduce a major issue we shall develop later: fire, that is, technology, is nothing less than an essential condition for the use of wisdom, which means not only that it is essential for human life but also that it is essentially human—technology is not a secondary ontological addition to human nature but a primary ontological quality, which defines human nature from the very beginning.

Going back to the three divine elements given to humans, there are at least two other interesting issues to reflect upon. The first is that technology is defined from the very beginning as an art (of course, according to the ancient meaning of the notion of art. See Tatarkiewicz 1980) founded on both practical and intellectual abilities. We may say that technology has been thought of from the very beginning as a result of the union between technique (as the practical way of mastering nature in a broad sense) and science (as the theoretical way of mastering nature in a broad sense). The second issue is that the divine foundation of technology leads us to think of it as a kind of double bridge, as it were. On the one hand, it is a kind of bridge between humans and nature, in that it establishes the kind of relationship that exists between them, in particular by making the former able to rule the latter. On the other hand, it is a kind of bridge between humans and the gods, in that it establishes the kind of relationship that exists between them, in particular by allowing the former to move toward the latter.

Humans' aspiration to move toward the gods is even further stressed by Ovid, who writes in his *Metamorphoses*:

> Still missing was a creature finer than these [than the animals], with a great mind, one who could rule the rest: man was born, whether fashioned from immortal seed by the Master Artisan who made this better world, or whether Earth, newly parted from Aether above, and still bearing some seeds of her cousin Sky, was mixed with rainwater by Titan Prometheus and moulded into the image of the omnipotent gods. And while other animals look on all fours at the ground he gave to humans an upturned face, and told them to lift their eyes to the stars. (Ovid *Met* I 77–87)

According to Ovid, Prometheus may be even the creator of humans, highlighting the technological characterization of humans as their primary ontological quality. More precisely, two details of Prometheus' method of creating humans are particularly illuminating. First, humans are "moulded into the image of the omnipotent gods": again, Prometheus' role is that of making humans analogous to the gods—we may say that technology's role, as the result of Prometheus' method of creation, is to make humans analogous to the gods. Second, Prometheus gives humans "an upturned face" and tells "them to lift their eyes to the stars." We may say, figuratively, that Prometheus gives humans a potential challenge and an actual challenge. As for the potential challenge we may read in the words "an upturned face," it distinguishes humans from any other animal (the "other animals look on all fours at the ground"): figuratively, what distinguishes humans is that, by nature, they aspire to something higher than the earth, that is, they do not settle for their present *status quo*, both in terms of their present quality of life and in terms of their quality of ontology. On the contrary, by their very nature, they aspire to exceed their present identity—Prometheus makes humans special earthly beings who, by nature, aspire to move from an earthly condition to a divine condition. As for the actual challenge we may read in the words "to lift their eyes to the stars," it is a kind of purpose in life to pursue: humans' "upturned face" has a precise aim, which is nothing less than "the stars," which, figuratively, are what most radically exceeds the earthly human ontology—paradoxically enough, Prometheus makes humans special earthly beings whose purpose in life to pursue is to exceed their ontology. And, as Prometheus is the god who creates humans as beings aspiring to most radically exceed their ontology, we may say that it is precisely technology, that is, the very result of Prometheus' method of creation, that allows humans to undertake the challenge.

More precisely, the partial result we get is the following: the kind of challenge that humans undertake through technology is changing and, more specifically, improving their ontology, by moving from a kind of earthly dimension to a kind of divine dimension (and this is precisely what distinguishes humans from any other animal).

Technology's essential role for humans' ontology is stressed by another detail of the myth of Prometheus: he is a Titan, and the Titans are the most ancient gods, predating even the Olympians ruled by Zeus (see Hes. *Theog.* 424, where they are described as πρότεροι θεοί). Thus, what is most illuminating is that, whoever is responsible for humans' technological ontology, whether as someone who creates them in a specific way or as someone who gives them specific qualities and tools, is so ancestral that we may legitimately think of technology not as something added to humans' essential ontology in their evolution over time but as something constituting humans' essential ontology from the very beginning—technology, that is, the quality of being technological, is an essentially human way of being.

And, as we have partly seen, being technological as an essentially human way of being defines what may be humans' greatest challenge, that is, challenging the divine. Other meaningful details of the myth of Prometheus help us understand this challenge. First, we should say that, according to Hesiod's *Works and Days*, following Prometheus' deceit, Zeus punishes humans not only by depriving them of fire but also by giving them Pandora, the first woman:

> And he called this woman Pandora, because all the gods who abide in Olympus gave her as a gift a pain for grain-eating men. But when the gods completed this deception of sheer doom, against which there is no remedy, father Zeus sent the famed Argos-killer to Epimetheus, the swift messenger of the gods, bringing the gift. Nor did Epimetheus take notice how Prometheus had told him never to accept a gift from Zeus the Olympian, but to send it right back, lest an evil thing happen to mortals. But he accepted it, and only then did he take note in his *noos* that he had an evil thing on his hands. (Hes. *Op.* 80–9)

Again, Epimetheus makes a mistake that Prometheus tries to remedy. But, in this case, he fails to succeed: Pandora got to humans and "took the great lid off the jar and scattered what was inside. She devised baneful anxieties for humankind. The only thing that stayed within the unbreakable contours of the jar was *Elpis* [Hope]. It did not fly out" (Hes. *Op.* 94–7). "Before this, the various kinds of humanity lived on earth without evils and without harsh labour, without wretched diseases that give disasters to men" (Hes. *Op.* 90–2). Pandora gives humans most interesting things: "pain," "baneful anxieties," "evils," "harsh labour," "wretched diseases," and "disasters"—precisely what technology can both relieve us of and afflict us with. Thus, even though Prometheus can do the most crucial things for humans, from improving to creating their lives, there is one thing he cannot do, at least directly: freeing their lives from "pain," "baneful anxieties," "evils," "harsh labour," "wretched diseases," and "disasters." It is no coincidence that this kind of ambiguity characterizes technology, too, which has an ambiguous relationship with humans (and this ambiguity is clearest in our technological era): on the one hand, technology is precisely what greatly improves human life and, on the other hand, technology is precisely what is a great danger to human life, being responsible for the evils that afflict it (in the myth, because of the consequences of Prometheus' actions and, in our technological era, because of the consequences of technology's extreme improvement). But what is most interesting to us is that there is an overlap between the evils that afflict human life and the evils that technology and, in particular, contemporary technology tries to remedy: for instance, physical "pain" relieved by an anesthetic, "baneful anxieties" relieved by anxiolytics, "harsh labour" relieved by a machine, "wretched diseases" relieved by surgeries, and "disasters" relieved by weather forecasts. We may go even further by saying that contemporary technology tries to radically remedy these evils: for instance, by trying to remove "pain," "baneful anxieties," and "wretched diseases" through the development of sophisticated bioengineering technologies; by trying to remove "harsh labour" through the development of sophisticated mechanical engineering technologies; and by trying to remove "disasters" through the development of sophisticated data science technologies.

However, what we may learn from the myth of Prometheus is that technology's ambiguity is constitutive, since humans themselves, who are essentially technological, are constitutively ambiguous. On the one hand, humans are ambiguous in that they are created by mixing both "some seeds of her cousin Sky" and the "Earth." Moreover, they are both "moulded into the image of the omnipotent gods" and living together with the "other animals." On the other hand, technology is both virtuous and vicious, in that it is both relieving and afflicting.

Second, we should remark that, according to Aeschylus' *Prometheus Bound*, Prometheus knows a secret unknown to Zeus about what will cause his defeat: Zeus' offspring will become more powerful than him and, finally, defeat him. Prometheus dialogues with Io:

> "But now no limit to my tribulations has been appointed until Zeus is hurled from his sovereignty." "What! Shall Zeus one day be hurled from his dominion? [. . .] By whom shall he be despoiled of the sceptre of his sovereignty?" "By himself and his own empty-headed purposes." "In what way? Oh tell me, if there be no harm in telling." "He shall make a marriage that shall one day cause him distress." "With a divinity or with a mortal? If it may be told, speak out." "Why ask with whom? I may not speak of this." "Is it by his consort that he shall be dethroned?" "Yes, since she shall bear a son mightier than his father." "And has he no means to avert this doom?" "No, none—except me, if I were released from bondage." (Aesch. *PB* 755–70)

The secret identity of Zeus' conqueror ("I may not speak of this") does not allow us to believe that there may be an overlap between their identity and humans. Yet, there is surely a correlation between Zeus' descendent destiny and humans' ascendant destiny: as the former's power decreases (he is going to be "hurled from his sovereignty"), the latter's power increases (see again Aesch. *PB* 436–70, 475–505). But what is most interesting to us is that there is surely an overlap between the one who is responsible for humans' technological ontology (Prometheus), the one who knows that someone is going to challenge and defeat Zeus (Prometheus), and the one who is going to be Zeus' last chance "to avert this doom" (Prometheus). If it is true that Prometheus is the symbol of the relationship between humans and technology, then it is also true that this coincidence is most meaningful: technology is both the cause of and the possible remedy for ("No, none—except me") the crisis of the balance between the human and the divine. More precisely, a decision can be made. On the one hand, there is the option of using technology in order to revolutionize the ontological hierarchy between the human and the divine by allowing the former to rule the latter (which, as we have seen, is something more constitutive of the human ontological identity). On the other hand, there may be an alternative use of technology (even though it is something less constitutive of the human ontological identity): choosing not to revolutionize the ontological hierarchy between the human and the divine, that is, choosing to use technology in order to improve the quality of human life without changing the quality of human ontology—technology may be used as what allows humans to improve their quality of life without changing their quality of ontology, that is, remaining human, even though greatly improved. However, we should stress the choice of the first option, which is something more constitutive of the human ontological identity, at least according to the myth of Prometheus, as well as according to what our technological era seems to show day by day. We may describe the consequence of the first choice as follows: the more technology develops, the more indiscernible the ontological difference between the human and the divine becomes (as we shall argue in the following pages).

The myth of Prometheus provides us with other meaningful details. As we have learned from both the etymology of the name "Prometheus" and the dialogue with Io told by Aeschylus in his *Prometheus Bound*, Prometheus has the ability to predict the future. Let us focus on the meanings of the most illuminating fact that the symbol of the relationship between humans and technology can predict the future. Let us try to identify what its most illuminating meanings may be:

1. One of technology's essential aims is to predict the future, in that technology, together with science on which it is founded[3], is what humans

use in order to know and, in particular, to try to control the future (which is a crucial issue for any human being, past, present, and future). More precisely, what humans try to do through technology is to make sure that what will happen in the future will be identical, or at least analogous, to what has happened in the past under human control (for instance, technology helps engineers ensure that the bridge they are going to build next month will work in the same way as the bridge they built last month and that it will keep doing so).

2. Moreover, the close relationship between technology and the prediction of the future is not limited to characterizing technology: it is characteristic of humans, in that it increasingly characterizes their ontology through the development of technology if it is true that the latter constitutes humans' essential ontology from the very beginning. More precisely, through the development of technology, humans become increasingly more divine by moving from a condition whose limits coincide with the knowledge of the past and the present to a condition whose limits extend to the knowledge of the (potentially unlimited) future—and, of course, extending human knowledge to the potentially unlimited future means becoming divine, that is, obtaining one of the most essential typically divine prerogatives of omniscience.

But the myth of Prometheus gives us one last detail, which may show a critical obstacle. Plato tells us that "although man acquired in this way the wisdom of daily life, civic wisdom he had not, since this was in the possession of Zeus" (Plat. *Prot.* 321 d). Indeed, even though Prometheus could steal "Hephaestus' fiery art and all Athena's" (Plat. *Prot.* 321 e), he "could not make so free as to enter the citadel which is the dwelling-place of Zeus, and moreover the guards of Zeus were terrible" (Plat. *Prot.* 321 d). Thus, humans have no "civic wisdom." We should specify that the words "civic wisdom" are translated by others as "political wisdom,"[4] which is a more literal translation since the Greek words are "τὴν δὲ πολιτικὴν." The choice of these words is most meaningful, since the notion of πόλις is one of the cornerstones of Greek culture: πόλις makes reference to the "city" not only in terms of an aggregation of people and buildings but also in terms of a model of government. More precisely, the notion of πόλις specifies that referring to "political wisdom" means referring to the ability to consider not only what is singular and individual but also, and especially, what is plural and common. As such, we may say that the myth of Prometheus makes us see humans' critical obstacle as their constitutive inability to use "political wisdom" when they increasingly develop as technology increasingly develops—humans are constitutively unable to consider what is plural and common and, in particular, to give it priority over what is singular and individual, when they become increasingly more powerful as technology becomes increasingly more powerful. And, of course, this is remarkably dangerous, as we shall see especially when we shall consider the modern Prometheus, as well as a possible contemporary Prometheus.

Indeed, the critical obstacle for humans showed by the myth of Prometheus is the most crucial danger of our technological era, as we shall see when we shall reflect upon algorithms: paradoxically enough, we are increasingly developing technologies that provide us with the divine prerogative of predicting the potentially unlimited future, but we are unable to predict their consequences on what is plural and common (i.e., humanity and, in particular, the future of humanity) by giving it priority over what is singular and individual (i.e., the extreme ambition and economic interest of a technology company, and even of the politicians of a state).

Thus, according to the myth of Prometheus, "political wisdom" is the divine prerogative that humans cannot obtain at all (which shall be a crucial issue for us). And, as we have seen, as Pandora opened the jar, out of which "baneful anxieties for humankind" flew, "the only thing that stayed within the unbreakable contours of the jar was *Elpis* [hope]. It did not fly out" (and we shall also reflect upon the possible meaning of *Elpis*).

THE MODERN PROMETHEUS

Even though the myth of Prometheus seems to surprisingly address typically contemporary issues, it represents the founding phase of the relationship between humans and technology during which the latter is what allows the former to survive and rule over nature. We can identify two other major phases of the relationship between humans and technology: the first phase has to do with what happened two centuries ago (represented by the modern Prometheus) and the second phase has to do with what is happening in our technological era (represented by a possible contemporary Prometheus). More precisely, as we shall see, the modern Prometheus represents the phase during which technology allows humans to challenge the divine by creating living human matter from nonliving matter, whereas a possible contemporary Prometheus represents the phase during which technology allows humans to challenge the divine by creating the divine itself.

The figure of the modern Prometheus was represented by authors working in the early decades of the nineteenth century: especially Mary Shelley, with her *Frankenstein. Or, The Modern Prometheus*, but also Percy Bysshe Shelley, among others.

Mary Shelley's *Frankenstein. Or, The Modern Prometheus* is, at least from our perspective, the most meaningful representation of the historical evolution of the figure of Prometheus as a powerful symbol of the modern relationship between humans and technology and fascinatingly, as well as instructively, shows us both what has changed and what has not changed at all since the birth of Prometheus in the Greek culture. Mary Shelley tells us the story of the modern Prometheus represented by the protagonist of her novel: Victor Frankenstein, a talented and ambitious scientist who, after his mother's death, challenges death itself, first, by trying to create a human being from corpses, that is, living human matter from nonliving matter, and, second, by trying to make his creature resistant to diseases and death itself. Frankenstein succeeds, but the creature terrifies him by being a kind of misshapen giant with superhuman resistance and strength. The creature is terrified by his reaction, too, and, after having stolen Frankenstein's diary, runs away. This is the beginning of a series of tragedies, both for the creature and for Frankenstein. The former, who is capable of goodness (he secretly helps a family of farmers), ends up killing after having been continuously rejected and persecuted as a monster. The latter feels extremely guilty and even ends up being one of his creature's victims (the creature kills Frankenstein's bride and indirectly causes both his father's death and his own death). When the creature and Frankenstein meet, the former makes the latter an offer: if Frankenstein will create an analogous woman, then the creature will disappear with her to a remote corner of the globe, no longer terrifying humans in society. Frankenstein starts to create the woman but begins to feel even guiltier, fearing that they, together with their possible offspring, may cause even greater harm to humans he wishes to protect. As such, he stops to create the woman and consequently suffers the creature's wrath. After Frankenstein's death, the creature commits suicide by setting fire to himself to make sure that it is not possible for others to understand and repeat what started his tragic life.

Let us start from the end. Mary Shelley chooses a meaningful ending using the most typically Promethean emblem, that is, fire, to stress the difference between ancient culture, for which fire represents construction, and modern culture, for which fire (also) represents destruction. More precisely, the fact that the creature commits suicide by setting fire to himself seems to mean the following:

1. The remedy for technology is technology (the former kind being the creature, as the symbol of modern technology, and the latter being fire, as the symbol of technology in general).
2. If something goes wrong with technology and, in particular, with its modern development, then the remedy is not at all, and cannot be, to avoid its use (again, technology is represented as essentially human and, therefore, unavoidable for humans: even a technological monster is human, in that he results from both corpses and the work of a scientist).

On the contrary, humanity's best remedy for technology is technology itself (the novel ends with the death of Frankenstein followed by the death of the creature, but fire keeps existing).

3. Technology can be remarkably destructive (as well as remarkably constructive, of course).

However, destruction may be used as a lesson not to avoid technology, as we have seen, but to stress mistakes and help us learn from them, so that we can start to develop technology in a more conscious way.

Moving from the end to the beginning of Mary Shelley's novel, we realize that a first major issue does not change at all if we compare the modern Prometheus with the ancient Prometheus: the reason why Frankenstein develops his most ambitious technology is precisely the Promethean creation of humans as special earthly beings with "an upturned face" who "lift their eyes to the stars." Indeed, the modern Prometheus is described as follows: "It was the secrets of heaven and earth that I desired to learn" (Shelley 1994: 36), not only because of his desire for knowledge but also because of his ambition, which leads Frankenstein to try to get the "glory [that] would attend the discovery if I could banish disease from the human frame and render man invulnerable to any but violent death!" (Shelley 1994: 37). Thus, the reason why the modern Prometheus develops his most ambitious technology does not change at all if we compare it with the ancient Prometheus' reason, in that they both want humans to be at least partly divine. In the myth, by giving humans divine elements in order to allow them to rule over nature. And in Mary Shelley's novel, by making humans, in the case of Frankenstein, able to learn anything ("It was the secrets of heaven and earth that I desired to learn") and, in the case of Frankenstein's creature, resistant to diseases and death itself ("I could banish disease from the human frame and render man invulnerable to any but violent death!").

But something goes wrong, as Frankenstein's desire for knowledge and ambition are so unlimited that they end up being dangerous: "After days and nights of incredible labour and fatigue, I succeeded in discovering the cause of generation and life; nay, more, I became myself capable of bestowing animation upon lifeless matter. But this discovery was so great and overwhelming that all the steps by which I had been progressively led to it were obliterated, and I beheld only the result" (Shelley 1994: 50). Moreover, "As the minuteness of the parts formed a great hindrance to my speed, I resolved, contrary to my first intention, to make the being of a gigantic stature" (Shelley 1994: 51). Indeed, Frankenstein's unlimited desire for knowledge and ambition end up causing him to run into epistemological mistakes: first, methodological inaccuracy ("all the steps [. . .] were obliterated"); second, methodological haste ("a great hindrance to my speed"); third, neglecting the ethical consequences of both methodological inaccuracy and methodological haste ("to make the being of a gigantic stature," which is the reason why he terrifies humans)—we may summarize by saying that the most important reason why Frankenstein's unlimited desire for knowledge and ambition end up being dangerous is that they prevent him from working on the possible negative, even irremediable, consequences of the kind of technology he is developing: indeed, "the disaster is irreparable" (Shelley 1994: 71).

We should stress the severity of the epistemological mistakes and, moreover, of their ethical consequences. The creature explains "how was I terrified when I viewed myself in a transparent pool! I was in reality the monster that I am, I was filled with the bitterest sensations of despondence and mortification. Alas! I did not yet entirely know the fatal effects of this miserable deformity" (Shelley 1994: 109). Thus, even though the creature (and, symbolically, technology) can be totally good ("I was benevolent and good," Shelley 1994: 96), the epistemological mistakes and their ethical consequences can make him (and, symbolically, technology) totally bad ("misery made me a fiend," Shelley 1994: 96). Yet, again, the remedy for technology is technology ("Make me happy [i.e., create an analogous woman], and I shall again be virtuous," Shelley 1994: 96). But the relationship between Frankenstein and the creature (and, symbolically, between humans and technology) is so imbalanced that its ethical consequences end up causing a kind of paradoxical overturning of its structure: the creature finally exclaims that "You are my creator, but I am your master: obey!" (Shelley 1994: 162). Most interestingly, the ultimate result is that what is created can get to dominate its creator—which is precisely the worst fear of our technological era, as we shall see: the more we can succeed in creating increasingly powerful technologies, the more they can end up being possibly out of our control, and even controlling us.

Paradoxically enough, technology can go wrong both because of failure and because of extreme success, at least in terms of being created as increasingly powerful and, therefore, possibly out of our control.

We have argued that a first major issue does not change at all if we compare the modern Prometheus with the ancient Prometheus. However, there is a meaningful detail that changes in what we have reflected upon: the ancient Prometheus is a Titan, that is, a divine being, whereas the modern Prometheus is the talented and ambitious scientist Frankenstein, that is, a human being. Thus, humans seem to start their existence as beings who can master technology by divine intervention and to evolve into beings who can master technology by themselves—the first meaningful detail that changes from the ancient Prometheus (as a Titan) to the modern Prometheus (as a scientist) is that humans seem to evolve into beings who replace the divine with themselves. However, what seems to remain a divine, rather than human, prerogative is "civic wisdom" or "political wisdom" as we have seen. On the one hand, when technology is made by divine intervention (or, we may say, when humans do not replace the divine with themselves), its outcomes are positive (humans survive and rule over nature). On the other hand, when technology is made by human intervention (or, we may say, when humans replace the divine with themselves, and, therefore, there is no "civic wisdom" or "political wisdom" at all), its outcomes are both positive and negative (humans can make a great scientific discovery without being able to use it for good).

Indeed, the major change between the ancient Prometheus and the modern Prometheus is the following: humans increasingly try to obtain a divine status—and, as we shall see, the major change between the modern Prometheus and a possible contemporary Prometheus is that humans increasingly try to even surpass a divine status.

First, as we have seen, the modern symbol of the relationship between humans and technology is a human being replacing a divine being. Second, what this human being does is obtaining the most typically divine prerogative: creating—more precisely, creating living human matter from nonliving matter. What Frankenstein does by "bestowing animation upon lifeless matter" is moving from a typically human ontology (i.e., being creatures) to a typically divine ontology (i.e., being creators). Indeed, this is a major step: Frankenstein is the symbol of that crucial step of the relationship between humans and technology in which the latter is the former's tool to obtain the most typically divine prerogative—and, therefore, to be divine, at least in terms of being creators (as we shall see, the last step of the relationship between humans and technology, that is, our technological era, adds other divine prerogatives). Thus, moving from antiquity to modernity, technology is humans' privileged way of trying not to retain their natural identity by surviving and ruling over nature but to change it by obtaining a first divine prerogative, that is, creating nature itself, and, in particular, human nature from nonliving matter.

There is another meaningful detail that changes moving from the ancient Prometheus to the modern Prometheus: even though both stories are structured around punishments, the punishments change in a meaningful way. More precisely, in the case of ancient culture, both Prometheus and humans are severely punished: the former is continuously tortured by an eagle that eats his liver, and the latter are deprived of fire. In both cases, Zeus is the punisher, which means that they are ruled by a divine dimension, superior to both technology and humans. As for the causes of the punishments, Prometheus is punished because he deceives Zeus and, moreover, because he helps humans compete with him, and humans are punished because they get closer to him. If we try to read the literary symbols, then we may say the following: if humans develop technology aimed at competing with the superior dimension that rules over them, then they end up being deprived of something essential, that is, their very condition for being capable of development. In the case of modern culture, again, both Prometheus (Frankenstein) and humans are severely punished: the former feels guilty, loses people closest to him, and ends up dying, and the latter become the creature's victims, terrorized and even killed by him. However, moving from antiquity to modernity, something important changes, since a third being is punished: the creature himself, that is, technology itself. More precisely, moving from antiquity to modernity, the punished are not only the providers of technology (Prometheus and Frankenstein) and the users of technology (humans) but also technology itself (the creature). This is an exceedingly important point. We have argued that the more technology develops, the more indiscernible the ontological difference between the human and the divine becomes. We can add that the more technology develops, the more indiscernible the ontological difference between the human and the technological becomes, too—and, as we shall see, these two points are exceedingly important if we try to argue that our technological era may be read as the most radical form of anarchism we have ever experienced. Indeed, the major difference between the ancient sequence of punishments and the modern sequence of punishments is the following:

1. In the first case, we have four different ontological dimensions: the divine as the Olympian (Zeus, the punisher); the divine as the Titan (Prometheus, the punished); the human (humans, the punished); and the technological (fire, neither punisher nor punished). Thus, we may summarize through three different ontological dimensions: those who punish, being therefore superior (the divine as the Olympian); those who are punished, being therefore inferior (the divine as the Titan and the human); and something that does not enter this ontological hierarchy (the technological).
2. In the second case, we have three different ontological dimensions: the human as the provider of technology (Frankenstein, punished and punisher as well, when he attacks the creature); the human as the user of technology (humans, punished and punishers as well, when they attack the creature); and the technological (the creature, punished and punisher as well, when he attacks Frankenstein and humans). Thus, we may summarize through one and the same ontological dimension, which is an exceedingly important point: the punishers and the punished coincide. The ontological differences, and the ontological hierarchy distinguishing the superior from the inferior, become indiscernible, and even dissolve, since the three different ontological dimensions end up having one and

the same ontological destiny—the divine dissolves into the human, and the human dissolves into the technological.

In particular, we have added that the more technology develops, the more indiscernible the ontological difference between the human and the technological becomes. Indeed, this is a major issue of our technological era, when we are reflecting upon the ontological status of the most sophisticated humanlike technologies: Frankenstein's creature, who is humanlike not only because he can live but also because he can be punished, shows one of the major issues of contemporary philosophy of technology, as we shall see—the more we move from antiquity to modernity and from modernity to our technological era, the more developing technology means humanizing the technological (and, at the same time, making the human divine, as we shall see).

Before moving from modernity to our technological era, let us mention Percy Bysshe Shelley's *Prometheus Unbound*, which is fairly illuminating if we reflect upon what changes from antiquity to modernity. The author stresses the difference between his Prometheus, who is symbolically unbound, and Aeschylus' Prometheus, who is symbolically bound and, moreover, who seems to reconcile with Zeus at the end of the trilogy we do not have at our disposal:

> The *Prometheus Bound* of Aeschylus supposed the reconciliation of Jupiter with his victim as the price of the disclosure of the danger threatened to his empire by the consummation of his marriage with Thetis. Thetis, according to this view of the subject, was given in marriage to Peleus, and Prometheus, by the permission of Jupiter, delivered from his captivity by Hercules. Had I framed my story on this model, I should have done no more than have attempted to restore the lost drama of Aeschylus; an ambition which, if my preference to this mode of treating the subject had incited me to cherish, the recollection of the high comparison such an attempt would challenge might well abate. But, in truth, I was averse from a catastrophe so feeble as that of reconciling the Champion with the Oppressor of mankind. The moral interest of the fable, which is so powerfully sustained by the sufferings and endurance of Prometheus, would be annihilated if we could conceive of him as unsaying his high language and quailing before his successful and perfidious adversary. (Shelley 1968: preface)

Percy Bysshe Shelley's Prometheus is not as powerful as Mary Shelley's Frankenstein, but his symbolic meaning is useful to understand the first historical turn of the Western notion of technology, that is, the transition from antiquity to modernity. The ancient Prometheus reconciles with Zeus by obeying him, which means that the ontological hierarchy distinguishing a superior dimension from an inferior dimension keeps existing: the former, being more powerful, keeps ruling, and the latter, being less powerful, keeps obeying. On the contrary, the modern Prometheus does not reconcile with, and does not obey, Zeus at all: the former is finally unbound, and the latter is finally dethroned, which means that the ontological hierarchy distinguishing a superior dimension from an inferior dimension does not keeps existing at all. More precisely, Prometheus' victory over Zeus translates into a victory for humans, the former being "the Champion [. . . of mankind" and the latter being "the Oppressor of mankind."

Thus, the balance between a superior ontological dimension (i.e., the divine, being more powerful and ruling) and an inferior ontological dimension (i.e., the human, being less powerful and obeying) is broken: as we shall see when we shall translate the Promethean literary symbols into strictly philosophical terms (especially in the second chapter), the first historical turn of the Western notion of technology means that, through technology, the human starts to dissolve its difference from the divine—in strictly philosophical terms, humans start to technologically dissolve the difference between the real (i.e., immanence, being limited and imperfect) and the ideal (i.e., transcendence, being unlimited and perfect).

A POSSIBLE CONTEMPORARY PROMETHEUS

It is surprising how both the ancient Prometheus and the modern Prometheus, respecting the very etymology of the name "Prometheus," can actually predict the future, that is, our technological era. Indeed, as we shall see (especially in the third chapter), we are developing technologies that are literally dissolving the divine into the human (as well as the human into the technological), at least in two meaningful ways:

1. Through technology, we are obtaining more and more typically divine prerogatives: creation, as well as omnipresence, omniscience, and omnipotence. This process is somehow predicted by the ancient Prometheus but explicitly starts in the nineteenth century with the modern Prometheus.
2. Through technology, we are doing something even more challenging: we are creating the divine itself—we are creating a technological entity characterized by the typically divine prerogatives of omnipresence, omniscience, and omnipotence. We are creating an immanent divine, which means, as I shall try to argue, that our technological era may be read as the most radical form of anarchism we have ever experienced.

I think that it is crucial to distinguish and, therefore, to understand, these two kinds of issues in order to read our technological era. If we summarize the symbolic meanings of the three major steps of the historical evolution of the figure of Prometheus, then we may say the following:

1. The first step (the ancient Prometheus) has to do with humans' technological ability to survive and rule over nature: the human seems to be able to be superior to the nature but does not seem to be able to be superior to the divine.
2. The second step (the modern Prometheus) has to do with humans' technological ability to challenge the divine, by obtaining the divine prerogative of creating: the human seems to be able to equal the divine.
3. The third step (a possible contemporary Prometheus) has to do with humans' technological ability to replace the divine, both by obtaining more and more divine prerogatives and, moreover, by creating the divine itself: the human seems to be able to be superior to the divine.

We shall reflect upon the third step especially in the third chapter, but let us take into account at least a case in point: when we are using our technological devices, from our smartphones to our computers, for instance in the case of Google searches, we are creating, through our data, a technological entity that is omnipresent (by being everywhere), omniscient (by knowing everything, especially about us), and omnipotent (by having power, especially over us). Google's algorithm is one of several analogous cases in which we are creating an immanent divine as a technological entity that is finally capable of mastering its creator through its omnipresence, omniscience, and omnipotence (which is precisely what the historical evolution of the figure of Prometheus can predict, especially in the case of Frankenstein mastered by his creature, who exclaims that "You are my creator, but I am your master: obey!"). For instance, as we shall see especially in the third chapter through analogous examples, Hewlett-Packard's algorithm ended up discriminating against blacks when using the webcam (see Leonard 2009), Nikon's algorithm ended up discriminating against Asians when using the camera (see Rose 2010), and Amazon's algorithm ended up discriminating against women when recruiting staff (see Dastin 2018).

Is there a possible contemporary Prometheus helping us understand our exceedingly complex technological era? Art, in general, and literature, in particular, seem to keep providing promising tools: "Now we're all God. Every one of us will soon be able to see, and cast judgment upon, every other. We'll see what He sees. We'll articulate His judgment. We'll channel His wrath and deliver His forgiveness. On a constant and global level. All religion has been waiting for this" (Eggers 2014: 398). These words are pronounced by a mysterious character (who studied divinity, worked as a computer scientist, and ended up falling into alcoholism) of Dave Eggers'

The Circle. A Novel. Eggers' possible contemporary Prometheus is, interestingly enough, a trinity, that is, the three creators of a technology company with the following aims. First, "We will become all-seeing" (Eggers 2014: 71) through sophisticated technologies that allow "Full transparency [that] would bring full access" (Eggers 2014: 470)—which means the creation of a kind of technological omnipresence. Second, "We will become [. . .] all-knowing" (Eggers 2014: 71) through sophisticated technologies that allow "no more not-knowing" (Eggers 2014: 470): "I had the idea that I would endeavour, in my lifetime, to know every person on Earth. Every person, even if just a little bit. To shake their hand or say hello. And when I had this inspiration, I really thought I could do it. Can you feel the appeal of a notion like that?" (Eggers 2014: 469)—which means the creation of a kind of technological omniscience. Third, "I'm a believer in the perfectibility of human beings. [...] I think we can be perfect or near to it. And when we become our best selves, the possibilities are endless. We can solve any problem. We can cure any disease, end hunger, everything, because we won't be dragged down by all our weaknesses, our petty secrets, our hoarding of information and knowledge. We will finally realize our potential" (Eggers 2014: 294)—which means the creation of a kind of technological omnipotence. However, one of the three creators realizes that there may be a danger in "This idea of Completion [...]. It has to be brought back into some kind of balance" (Eggers 2014: 485), since "Completion is the end. We're closing the circle around everyone—it's a totalitarian nightmare" (Eggers 2014: 486). More precisely, he thinks that "Completion," which we may define as a creation of an omnipresent, omniscient, and omnipotent technological entity, is a kind of "Infocommunism [. . .] paired with ruthless capitalistic ambition" (Eggers 2014: 489), which is created by someone who "professionalized our idealism, monetized our utopia" (Eggers 2014: 489). The dystopia described in Eggers' novel may bring to mind an aspect of our technological era. "Already, 90 percent of the world's searches go through the Circle. Without competitors, this will increase. Soon it'll be nearly 100 percent. Now, you and I both know that if you can control the flow of information, you can control everything. You can control most of what anyone sees and knows. If you want to bury some piece of information, permanently, that's two seconds' work. If you want to ruin anyone, that's five minutes' work. How can anyone rise up against the Circle if they control all the information and access to it? They want everyone to have a Circle account, and they're well on their way to making it illegal not to. What happens then? What happens when they control all searches, and have full access to all data about every person? When they know every move everyone makes? If all monetary transactions, all health and DNA information, every piece of one's life, good or bad, when every word uttered flows through one channel?" (Eggers 2014: 487–8). "Everyone will be tracked, cradle to grave, with no possibility of escape" (Eggers 2014: 486). Thus, the solution proposed by the creator is to establish "The Rights of Humans in a Digital Age," which are the following: "We must all have the right to anonymity. Not every human activity can be measured. The ceaseless pursuit of data to quantify the value of any endeavour is catastrophic to true understanding. The barrier between public and private must remain unbreachable. We must all have the right to disappear" (Eggers 2014: 490). But the novel ends differently: the creator's proposal ends up being unheard, and the dystopia keeps worsening.

Trying to read an exceedingly complex phenomenon, especially when it is ongoing, is complicated. As such, we cannot say that the trinity of the creators described in Eggers' novel makes for the best possible contemporary Prometheus. However, we can say that it may be a possible contemporary Prometheus who is able to show us that the second, and the last, historical turn of the Western notion of technology means not only that, through technology, the human dissolves its difference from the divine but also that the human even surpasses the divine—the human even surpasses the divine by technologically creating it.

And, as we have seen, the technological becomes humanized. Our technological era is actually characterized by the kinds of questions that the modern Prometheus (Frankenstein) starts to ask by creating a technology (the creature) that obtains typically human prerogatives, such as autonomously feeling, thinking, and acting (see at least Coeckelbergh 2010; Floridi 2008). There are several analogous examples in contemporary art: for instance, Ian McEwan's *Machines Like Me and People Like You*. But they are extreme: they show that the typically human prerogative that contemporary technology is obtaining is making decisions—more precisely, making decisions for humans (as we shall see especially in the third chapter). McEwan's novel starts with a reference to Mary Shelley: an important character says that she would be glad to share with her the technological experience she is going to have, which is interacting with an extremely humanized machine meaningfully called Adam. Also, the issue of trying to obtain divine prerogatives shows up immediately: the protagonist, who is also the narrator, says that creating Adam means trying to avoid the mortal condition. Moreover, the issue of technology obtaining human prerogatives is even clearer: "[Adam] was sentient. He had a self. How it's produced, wet neurons, microprocessors, DNA networks, it doesn't matter. Do you think we're alone with our special gift?" (McEwan 2019: 303–4). But what is clearest is another essential issue (which we shall reflect upon in the fourth chapter): the relationship between artificial intelligence and the human mind and, in particular, the question of their possible hierarchy. McEwan stresses the essentiality of this issue by having it introduced by Turing, who shows up as a character dialoguing with the protagonist. More precisely, Turing says:

> Have you any idea what it takes to catch a ball, or raise a cup to your lips, or make immediate sense of a word, a phrase or an ambiguous sentence? We didn't, not at first. Solving maths problems is the tiniest fraction of what human intelligence does. We learned from a new angle just how wondrous a thing the brain is. A one-litre, liquid-cooled, three-dimensional computer. Unbelievable processing power, unbelievably compressed, unbelievable energy efficiency, no overheating. The whole thing running on twenty-five watts—one dim light bulb. But so far science has had nothing but trouble understanding the mind. Singly, or minds *en masse*. The mind in science has been little more than a fashion parade. Freud, behaviourism, cognitive psychology. Scraps of insight. Nothing deep or predictive that could give psychoanalysis or economics a good name. So, knowing not much about the mind, you want to embody an artificial one in social life. Machine learning can only take you so far. You'll need to give this mind some rules to live by. How about a prohibition against lying? According to the *Old Testament*, *Proverbs*, I think, it's an abomination to God. But social life teems with harmless or even helpful untruths. How do you separate them out? Who's going to write the algorithm for the little white lie that spares the blushes of a friend? Or the lie that sends a rapist to prison who'd otherwise go free? We don't yet know how to teach machines to lie. And what about revenge? Permissible sometimes, according to you, if you love the person who's exacting it. Never, according to your Adam. (McEwan 2019: 302–3)

As we shall see in the fourth chapter, this is an essential issue. I shall try to address it by starting with the following provocative question: is it actually intelligent to devote all our energy to relying on artificial intelligence resulting from the externalization of a mere fraction of what happens in our mysterious, complex, and surprisingly is refered to powerful! minds? And is it actually intelligent to devote almost none of our energy to discovering what our mysterious, complex, and surprisingly powerful minds can do, even in addition to intelligence itself, which is only one of our minds' abilities? These questions are clearly rhetorical: I shall try to argue the reasons why their answers are negative.

As for a possible contemporary Prometheus, I should end the first chapter by saying that both the novels I have been referring to do not have a happy ending. In Eggers' novel, the dystopia even worsens: the typically human prerogatives disappear almost totally. In McEwan's novel, the protagonist, feeling guilty, ends up admitting his failure: his use of technology has been the worst, and the ultimate result is that technology has been destroyed, and his life has worsened. The absence of a happy ending is something that a possible contemporary Prometheus shares with the modern Prometheus, whose story ends in a tragic way. However, the birth of Prometheus in the Greek culture seems to suggest a possible solution: even though the myth of the ancient Prometheus ends with Pandora's tragic opening of the jar, "the only thing that stayed within the unbreakable contours of the jar was *Elpis* [Hope]. It did not fly out." Thus, even though "she devised baneful anxieties for humankind," "evils," "harsh labour," and "wretched diseases that give disasters to men," humans maintain a possible solution, that is, a hope yet to be discovered. Moreover, we should not forget that, if it is true that the historical evolution of the figure of Prometheus represents humans as trying to obtain typically divine prerogatives, then it is also true that one divine prerogative is missing: "civic wisdom he had not, since this was in the possession of Zeus." As I shall try to argue, "civic wisdom" may constitute humanity's hope—humanity's hope to also obtain the crucial divine prerogative of being able to develop technology in a "civic" way, that is, by considering what is plural and common and, in particular, by giving it priority over what is singular and individual, both synchronically and diachronically.

In any case, what any Prometheus shows us is that humanity's historical evolution coincides with trying, through technology, to move from a human ontology to a divine ontology—and the last step of humanity's historical evolution seems to coincide with the creation of an immanent divine as a technological entity, which means, as I shall try to argue in the second chapter, that our technological era may be read as the most radical form of anarchism we have ever experienced.

NOTES

1. There are several references. See at least Dougherty (2006).
2. See https://www.sens.org/: "SENS Research Foundation works to develop, promote, and ensure widespread access to therapies that cure and prevent the diseases and disabilities of aging by comprehensively repairing the damage that builds up in our bodies over time. We are redefining the way the world researches and treats age-related ill health, while inspiring the next generation of biomedical scientists."
3. On the relationship between technology and science see at least Arthur (2011) and Latour (1987).
4. See, for instance, the translation by Jowett, revised by Ostwald.

Chapter 2

Homo anarchicus

THE ANCIENT INVENTION OF IDEALITY AND ITS TECHNOLOGICAL MEANING

What I shall try to argue in the second chapter is that, interestingly enough, the three major steps of the historical evolution of the figure of Prometheus correspond to the three major steps of the historical evolution of the foundation of the Western culture itself from an epistemological perspective: the invention of ideality and its relationship with reality (see at least Berlin 1988; Jullien 2009. I have explored this issue extensively, especially in Chiodo 2013, 2014, 2015, 2016, 2018). More precisely, the following can be said of the Western culture:

1. It has been founded on the invention of the dimension of ideality as the perfect counterpart of the imperfect dimension of reality—the former being complete, invariable, and universal and the latter being incomplete, variable, and particular. Indeed, the invention of the dimension of ideality may be thought of as the most defining feature distinguishing the Western culture from other cultures, at least from an epistemological perspective (it is no coincidence that François Jullien, when comparing the European culture with the Chinese culture, writes that what distinguishes the former is not a religious issue but the dimension of ideality, which is totally absent from the latter. See Jullien 2009). More precisely, the synergy between the Greek culture, in general, and the Greek philosophy, in particular, on the one hand, and the move from polytheisms to monotheisms, on the other hand, have defined the Western culture as distinguished by a precise epistemological structure: reality's quest for

ideality—the latter being the former's model in all cultural spheres, from science and technology to art, from law to religion (see Chiodo 2018).
2. It has experienced a crisis of the relationship between reality and ideality, by moving from their epistemological dualism to its removal: after millennia in which ideality has been reality's asymptotic model, that is, the best tool to develop its (incomplete, variable, and particular) ontology, since the first decades of the nineteenth century, ideality has been thought of not as reality's best tool to develop its ontology but as reality's ontological destiny, by trying to move from imperfect incompleteness, variability, and particularity to perfect completeness, invariability, and universality (see at least Berlin 1999, 2006).
3. It has experienced a tragic failure, after having tried to make the ideal real, especially through radical forms of totalitarianism, from epistemology to ethics to politics. And, after the crisis and failure of the relationship between reality and ideality, the latter has started to weaken and the former has started to make itself absolute, that is, the unique dimension: this is the beginning of forms of anarchism, from epistemology to ethics to politics (see at least Feyerabend 1975; Rorty 1979, 1989)—more precisely, this is the beginning of the most radical form of anarchism that, as I shall try to argue, may be thought of as the most defining feature distinguishing of our technological era.

I believe that the three epistemological turns may be surprisingly promising as a way of understanding both the meaning of technology in the Western culture, in general, and our technological era, in particular, since they actually define what technology is for humans, and, therefore, who humans are, and will become, through technology.

As we have learned from the figure of Prometheus, technology is essentially human, in that it is not a secondary ontological addition to human nature but a primary ontological quality that has been defining human nature from the very beginning. However, what changes throughout the Western history is the kind of ontological status obtained by humans through their constitutive use of technology by developing it through three phases that correspond to the three major steps of the historical evolution of the figure of Prometheus and, at the same time, to the three epistemological turns we have introduced.

Let us start from the first epistemological phase, which is founded in a systematic way by Plato's philosophy. In particular, let us carefully read what Plato writes in his *Republic*:

> "In the present case, then, let us take any multiplicity you please; for example, there are many couches and tables." "Of course." "But these utensils imply, I suppose, only two ideas or forms, one of a couch and one of a table." "Yes."

> "And are we not also in the habit of saying that the craftsman who produces either of them fixes his eyes on the idea or form, and so makes in the one case the couches and in the other the tables that we use, and similarly of other things? For surely no craftsman makes the idea itself. How could he?" "By no means." (Plat. *Resp.* 596 a–b)

These words are crucial: Plato uses them as the basis upon which to establish an analogy between the craftsman and the Demiurge, that is, the ultimate creator. Thus, these words describe not a method but the method of creation for anything, from the Demiurge's earth to the craftsman's couches and tables—Plato's words describe the working method of technology and, in particular, the classical working method of technology, which lasted for several centuries. More precisely, if the craftsman wants to make real, that is, particular couches and tables, then he should make reference to ideal, that is, universal couches and tables, which work as ideal models—what Plato argues is that, if the craftsman wants to make real couches and tables that are good, then he should refer to ideal couches and tables since, even though his real couches and tables, being particular, will necessarily be imperfect, incomplete, and variable, the most promising working method to make them as good as possible is to refer to ideal couches and tables, being universal, that is, perfect, complete, and invariable. Ideality's perfection, working as reality's ideal model, is the most promising working method for making something real and, in particular, real artifacts, by guiding reality toward a progressive improvement of itself, and, in particular, by moving reality from a more imperfect ontological status to a less imperfect ontological status. Thus, the craftsman "fixes his eyes on the idea or form, and so makes in the one case the couches and in the other the tables that we use," that is, his working method coincides with using ideality's perfection as reality's ideal model to make it as good as possible and, in particular, to make it increasingly better.

However, even though real artifacts may be as good as possible and increasingly improve, they cannot change their ontological status: real artifacts are necessarily imperfect, that is, particular, incomplete, and variable. Plato masterfully explains that it is logically impossible to make perfect, that is, universal, complete, and invariable, real artifacts—it is logically impossible to make the ideal real. Saying "'For surely no craftsman makes the idea itself. How could he?' 'By no means'" means the following: the idea, that is, the ideal model, is a kind of universal ideal matrix for any particular real artifact. Being a kind of universal ideal matrix means being the ideal rule for what any particular real artifact should be essentially in order to be a good artifact of its kind. For instance, the idea of a couch guides any real couch toward fitting the human measure in terms of shape, size, comfort, etc. Let us imagine that we are the craftsman who is going to make a real couch. Even though we can think of the idea of a couch as the ideal rule of what our couch should essentially be in order to be a good couch in terms of shape, size, comfort, etc., when we move from thinking to making, we realize that we must necessarily particularize. We must necessarily choose a particular shape (rounded, for instance), excluding all others (squared, for instance). We must necessarily choose a particular size (two meters, for instance), excluding all others (three meters, for instance). We must necessarily choose a particular way of making our couch comfortable (using leather, for instance), excluding all others (using velvet, for instance). And particularizing means moving from the ontological status of ideality, which is universal, coinciding with universal rules of perfection, completeness, and invariability to the ontological status of reality, which is particular, coinciding with particular artifacts that, resulting from particular choices, are particular, imperfect, incomplete, and variable, since they necessarily have alternatives that may be better. As such, we may say that what Plato masterfully explains is that ontologies of rules and ontologies of applications are necessarily different—we may say that the reason why it is logically impossible to make the ideal real is that the ideal rule, which is universal, cannot be its real application, which is particular.

Yet, making reference to ideal rules is the most promising working method invented by the Western culture to make good real applications and, in particular, to make them increasingly improve. And the dualism between ideal rules and real applications is not only a logical necessity but also a great advantage. If the craftsman makes his couch by referring to a real application, which is another real couch, then there are at least two disadvantages. First, he refers to something that is necessarily imperfect, incomplete, and variable: the model he uses, being real and particular, cannot allow him to choose a particular shape, a particular size, and a particular comfort that most promisingly result from their comparisons with perfection itself. In particular, he is likely to refer to the real couch that is typically used where he lives to make an almost identical couch and to keep making almost identical couches for the rest of his life. The ultimate result is that the real couches he makes are likely to be incapable of progressive improvement.

Second, he loses the most important ability: the ability of infinite progressive improvement, which is something that only an ideal model, that is, an asymptotic model, can provide. If I am the craftsman and my model is a real couch, which is particular and, therefore, imperfect, then I have an actual chance of making a couch that is better than my model. But, if I am the craftsman and my model is an ideal couch, which is universal, and, therefore, perfect, then I have no actual chance at all of making a couch that is better than my model. My model will keep guiding me (by inspiring me as well as challenging me) for the rest of my life, and I will keep trying to increasingly improve my couches—the most important possibility that an ideal model gives us is that, by being asymptotic, it pushes us to asymptotically improve our artifacts, which means that we will infinitely work in an effort to improve our artifacts and, finally, to develop our abilities.

Thus, Plato's masterful lesson may be summarized as follows: about two and a half millennia ago, we, as Western humans, invented a special way of ruling over nature, in general, and developing technology, in particular—we invented ideality as the perfect counterpart of reality, that is, as the model of reality and, therefore, as the working method of technology: if we want to make an artifact that is as good as possible and increasingly improves, then we should refer to its ideal model.

This working method lasted for several centuries, and Immanuel Kant gives us other meaningful insights in his *Critique of Pure Reason*, especially by distinguishing the notion of ideal from the notion of idea, the latter resulting from Plato's philosophy and the former resulting, at least in a systematic way, from Kant's philosophy. According to Kant, the dualism between ideality and reality is strengthened by the notion of ideal: "something that seems to be even further removed from objective reality than the idea is what I call the ideal, by which I understand the idea not merely *in concreto* but *in individuo*, that is, as an individual thing which is determinable, or even determined, through the idea alone" (Kant 1781: A 568/B 596). For instance, "Virtue, and with it human wisdom in its entire purity, are ideas. But the sage (of the Stoics) is an ideal, i.e., a human being who exists merely in thoughts, but who is fully congruent with the idea of wisdom" (Kant 1781: A 569/B 597). In particular, "just as the idea gives the rule, so the ideal in such a case serves as the original image for the thoroughgoing determination of the copy" (Kant 1781: A 569/B 597), which means that "we have in us no other standard for our actions than the conduct of this divine human being, with which we can compare ourselves, judging ourselves and thereby improving ourselves, even though we can never reach the standard" (Kant 1781: A 569/B 597). Thus, the ideals "provide an indispensable standard for reason, which needs the concept of that which is entirely complete in its kind, in order to assess and measure the degree and the defects of what is incomplete" (Kant 1781: A 569/B 597-A 570/B 598), and "have a practical power (as regulative principles) grounding the possibility of the perfection of certain actions" (Kant 1781: A 569/B 597). We may summarize by saying that from Plato to Kant, as well as from the notion of idea to the notion of ideal, ideality as the perfect counterpart of reality has strengthened the following:

1. The dualism that distinguishes it from reality.
2. Consequently, its power to be the most promising working method invented by the Western culture to make reality increasingly improve

(in Kant's words, first, by giving us the "standard for our actions," "with which we can compare ourselves, judging ourselves," and, second, by making us "improv[e]").

Whereas Kant focuses on improving human actions in general, that is, human ethics, his insights may also apply to human artifacts in particular, at least from our perspective: the ideal may be also the "standard" that gives us "the concept of that which is entirely complete in its kind, in order to assess and measure the degree and the defects of what is incomplete," which may be also the artifact we make. When Kant explains the genesis of the ideal in his *Critique of the Power of Judgment*, other important insights show up. Whereas Plato's idea predates humans, the genesis of Kant's ideal is given by "the aesthetic normal idea, which is an individual intuition (of the imagination) that represents the standard for judging it as a thing belonging to a particular species" (Kant 1790: 5, 233). Through "the aesthetic normal idea," it is possible "to superimpose one image on another and by means of the congruence of several of the same kind to arrive at a mean that can serve them all as a common measure" (Kant 1790: 5, 234). For instance, in the case of the ideal of "beautiful man" (Kant 1790: 5, 234), "if in a similar way there is sought for this average man the average head, the average nose, etc., then this shape is the basis for the normal idea of the beautiful man in the country where this comparison is made" (Kant 1790: 5, 234). The genesis of the ideal requires another essential element in addition to "the aesthetic normal idea": "the idea of reason, which makes the ends of humanity insofar as they cannot be sensibly represented into the principle for the judging of its figure, through which, as they effect in appearance, the former are revealed" (Kant 1790: 5, 233), which means that the ideal of "beautiful man" requires "the expression of the moral, without which the object would not please universally and moreover positively" (Kant 1790: 5, 235)—the ideal requires "the visible expression of moral ideas" (Kant 1790: 5, 235).

As we have seen, Kant's focus is the human being, whose ideal results from the following:

1. Something provided by human imagination, that is, "the aesthetic normal idea," which, starting from our empirical experience, gives us "a mean," "a common measure" of what exists.
2. Something provided by human reason, that is, "the idea of reason," which gives us something more "universal," being "the moral"—and "the moral" is precisely what makes the ideals able to "provide an indispensable standard for reason, which needs the concept of that which is entirely complete in its kind, in order to assess and measure the degree and the defects of what is incomplete," and to "have a practical power

(as regulative principles) grounding the possibility of the perfection of certain actions." Thus, the ideal makes us compare what is imperfect with what is perfect and, therefore, makes us perfect the former through the latter, even though asymptotically—again, the ideal makes us increasingly improve the real, even though asymptotically (Kant defines the ideal not only, as we have seen, as "even further removed from objective reality," existing "merely in thoughts" and working as "the standard" "we can never reach", but also as follows: "one may never concede them [the ideals] objective reality (existence)" since real things "always fall infinitely short of reaching," Kant 1781: A 569/B 597 and A 578/B 606).

We may see in Kant's words the epistemological structure underpinning the most promising working method invented by the Western culture to make good real applications and, in particular, to make them increasingly improve. More precisely, we may correlate Kant's two elements underpinning the ideal with the two epistemological phases underpinning technology's working method: abstraction and idealization.

Let us offer an example by going back to Plato's craftsman. Both Plato and Kant think of the dimension of ideality as the perfect counterpart of the dimension of reality. However, whereas Plato's idea predates humans, Kant's ideal may be thought of as follows:

1. On the one hand, human imagination (through "the aesthetic normal idea") can, first, "superimpose" the images of all the objects empirically experienced by someone and, second, find their "mean," their "common measure." If we try to move from Kant's focus on the human being to a possible focus on technology by going back to the simplest example of the craftsman, then we may say that his imagination can, first, compare the images of all the couches empirically experienced by him and, second, identify what they share (their "mean," their "common measure") by distinguishing it from what they do not share. For instance, the craftsman ends up thinking that the couches share a shape that is made in order to fit the measure of a human body. Thus, fitting the measure of a human body is a quality that a couch, that is, any couch, should have in order to be a good couch. However, this is untrue, as couches do not share the same shape in order to fit the measure of a human body: sometimes they do so by being rounded and sometimes they do so by being squared. Indeed, speaking of identification of what they share by distinguishing it from what they do not share means speaking of abstraction, which is an epistemological process that is essential to the Western culture, clearly defined both by John Locke as what human "understanding stores away for use as standards" (Locke 1690: II,

11, 9),[1] after having "retain[ed] only what is common to them all [the objects experienced]" (Locke 1690: III, 3, 7),[2] and by contemporary philosophers as "the omitting of properties" (Nowak 1990: 193) not shared by the objects experienced and, therefore, "omission of a truth" (Jones 2005: 175) and "omission without misrepresentation" (Jones 2005: 185).

2. On the other hand, human reason (through "the idea of reason") can perfect what results from human imagination. If we try to move from Kant's focus on the human being to a possible focus on technology by going back to the simplest example of the craftsman, then we may say that his reason can get to the ideal model of the couch, that is, any couch, by definitely moving from real couches to their ideal rules. For instance, the craftsman ends up thinking that the ideal model of the couch requires that a couch, that is, any couch, should fit the human measure in terms of shape, size, and comfort in order to be a good couch. But it is unlikely that all the couches empirically experienced by him could fit all three rules: sometimes they may sufficiently fit the first rule, but scarcely fit the others; sometimes they may sufficiently fit the second rule, but scarcely fit the others; sometimes they may sufficiently fit the third rule, but scarcely fit the others, etc. And the couch he is going to make, no matter how good it may be, cannot be as perfect as its ideal model, since the former is a particular couch and the latter is its universal set of rules. Indeed, speaking of getting to the ideal model by definitely moving from real couches to their ideal rules means idealizing, which is an epistemological process that is even more essential for the Western culture, clearly defined by contemporary philosophers as "assertion of a falsehood" (Jones 2005: 175) and "misrepresentation" (Jones 2005: 185).

Thus, by abstracting, we identify what real objects share and, by idealizing, we get to their ideal model—moreover, we think that there is a rigorous epistemological dualism between real objects and their ideal model, since the former are particular objects and the latter is their universal set of rules.

This is how we, as Western humans, have been thinking for millennia, working on the foundation not only of our technology but also of our science (which underpins technology itself),[3] art, law, and religion—in all spheres of our culture, ideal models guide us, being the most defining feature distinguishing the Western culture from other cultures.[4]

If we go back to the ancient Prometheus, then we can identify an essential correlation between the kind of relationship between humans and technology symbolized and the epistemological dualism between real objects and their ideal model. The ancient Prometheus symbolizes the following kind of relationship translated into epistemological terms:

1. Humans, without something ontologically superior and divine (the Gods' fire and wisdom in the arts), that is, technology, are "witless," see "to no avail," do "not understand," dwell "beneath the ground like swarming ants, in sunless caves," and manage "everything without judgment." On the contrary, humans, with technology, "have sense," "eyes to see," "ears," "knowledge of houses built of bricks and" turn "to face the sun," and learn "to discern." Moreover, humans, with technology, know "numbers, too, chiefest of sciences," "and the combining of letters." In epistemological terms, we can say that without something ontologically superior, that is, ideality, reality is worse, in that it is imperfect, incomplete, and variable. On the contrary, with ideality, reality is asymptotically better, being asymptotically improving. In Plato's words, the craftsman "fixes his eyes on the idea" to make his couch as good as possible. In Kant's words, "we have in us no other standard for our actions than the conduct of this divine human being, with which we can compare ourselves, judging ourselves and thereby improving ourselves."
2. Humans challenge the divine's superior ontology: indeed, they not only rule over nature but also have "an upturned face" and "lift their eyes to the stars." However, humans do not obtain the divine's ontology at all: indeed, they remain ontologically human, that is, they have severe limits ("pain," "baneful anxieties," "evils," "harsh labour," "wretched diseases," and "disasters"), lack divine prerogatives (for instance, "civic wisdom he [man] had not"), and are earthly ("wretched mortal men, who dwell upon the earth"). Moreover, as Percy Bysshe Shelley highlights, "The *Prometheus Bound* of Aeschylus supposed the reconciliation of Jupiter with his victim," which means that, no matter how humans challenge the divine, the former remain ontologically human and, therefore, inferior, and the latter remains ontologically divine and, therefore, superior. In epistemological terms, we can say that reality challenges ideality's superior ontology: indeed, the former makes the latter its model and, therefore, increasingly improves. However, reality does not obtain ideality's ontology at all: indeed, it remains something whose progressive improvement is asymptotic, that is, unable to obtain perfection, completeness, and invariability. In Plato's words, "for surely no craftsman makes the idea itself." In Kant's words, "one may never concede them [the ideals] objective reality (existence)" since real things "always fall infinitely short of reaching."

We can summarize the meaning of this essential correlation by saying that we, as Western humans, have been thinking for millennia that technology is something most special, allowing us to increasingly improve reality and ourselves by comparison with their ideal models but through which we cannot totally dissolve reality and ourselves into their ideal models—we have been thinking for millennia that technology is the most special way of making reality and ourselves as good as possible but not of totally changing the very essence of their ontological status, which remains imperfect, incomplete, and variable and, therefore, asymptotically aspiring and, consequently, infinitely improving.

THE MODERN CRISIS OF IDEALITY AND ITS TECHNOLOGICAL MEANING

Most interestingly, when Mary Shelley and Percy Bysshe Shelley were writing their modern versions of the figure of Prometheus, the Western culture was experiencing its first epistemological revolution: the crisis of the relationship between reality and ideality, by moving from their epistemological dualism to its removal, that is, from thinking of ideality as reality's asymptotic model to thinking of ideality as reality's ontological destiny.

Isaiah Berlin, who has extensively worked on this issue (see at least Berlin 1988, 1990, 1996, 1999, 2000, 2006), gives us a useful insight into understanding the meaning of this epistemological revolution by highlighting the words chosen by Aleksandr I. Herzen (see Herzen 1956) to describe the first decades of the nineteenth century as characterized by a new kind of human sacrifice: the sacrifice "of living human beings on the altars of abstractions—nation, Church, party, class, progress, the forces of history: if these demand the slaughter of living human beings, they must be satisfied" (Berlin 1990: 16). In strictly philosophical terms, Berlin is speaking not of abstractions but of ideals, that is, the ideal models of reality, in the following terms: politics (through the ideal models of "nation" and "party"), religion (through the ideal model of "Church"), society (through the ideal model of "class"), economics (through the ideal model of "progress"), and history (through the ideal model of "the forces of history").

In particular, the first decades of the nineteenth century are characterized by idealistic philosophy, which revolutionizes the classical epistemological relationship between ideality and reality in general and the classical working method of technology in particular (as Mary Shelley's and Percy Bysshe Shelley's modern versions of the figure of Prometheus show). In particular, Georg W. F. Hegel revolutionizes the notion of ideal (first, in his aesthetics and, second, in his philosophy of history, by translating the epistemological structure of the former into the epistemological structure of the latter): "the Idea as the beauty of art is [. . .] the Idea as shaped forward into reality and as having advanced to immediate unity and correspondence with this reality" (Hegel 1975: *Introduction*, 8, i). "Taken thus, the Idea as reality, shaped in accordance with the Concept of the Idea, is the *Ideal*" (Hegel 1975:

Introduction, 8, i). "Accordingly there is here expressed the demand that the Idea and its configuration as a concrete reality shall be made completely adequate to one another" (Hegel 1975: *Introduction*, 8, i). The epistemological revolution is clear: according to Hegel, what distinguishes the ideal from the idea is its total embodiment into the real by totally coinciding with it.

The embodiment of ideality into reality is particularly clear in aesthetics, where the work of art at its best is the ideal by being something real that embodies the idea. For instance, "Raphael's Madonnas [. . .] show us forms of expression, cheeks, eyes, nose, mouth, which, as forms, are appropriate to the radiance, joy, piety, and also the humility of a mother's love" (Hegel 1975: III, A, 1). Thus, "the nature of the artistic Ideal is to be sought in this reconveyance of external existence into the spiritual realm, so that the external appearance, by being adequate to the spirit, is the revelation thereof" (Hegel 1975: III, A, 1).

And the coincidence between the "external existence" and "the spiritual realm," that is, between reality and ideality, is the epistemological structure that aesthetics and philosophy of history share. In philosophy of history, the embodiment of ideality into reality becomes crucial. In particular, there is a crucial analogy between the ideal as the work of art at its best, created by a human being, and the ideal as the monarch of the state, a human being themselves. Hegel's words to describe the former and the latter are analogous: "The personality of the state has actuality only as a person, as the monarch. Personality expresses the concept as such, whereas the person also embodies the actuality of the concept, and only when it is determined in this way [i.e., as a person] is the concept Idea or truth" (Hegel 1991: 317). And believing that the monarch of the state embodies the ideal means believing that, if it is true that they embody the perfect ideal, then it is also true that they must rule over everyone to guide them toward perfection—believing that the ideal can be made real means underpinning the theoretical condition for epistemological totalitarianism in general (and believing that the ideal can be made real by the monarch of the state means underpinning the theoretical condition for political totalitarianism in particular).

This epistemological revolution, which is particularly clear in Hegel's idealistic philosophy, characterizes romanticism in general, as well as other idealistic philosophical works. In particular, according to the romantics, everyone's life must embody their own ideals, as Friedrich Schiller expresses in his poem *The Ideal and the Actual Life*: "When, through dead stone to breathe a soul of light,/With the dull matter to unite/The kindling genius, some great sculptor glows;/Behold him straining, every nerve intent/Behold how, o'er the subject element,/The stately thought its march laborious goes!" (Schiller 1844: 171). Interestingly enough, Schiller explicitly refers to Johann G. Fichte's idealistic philosophy and, in particular, to his *Some Lectures*

Concerning the Scholar's Vocation, which marks a kind of move from Kant's philosophy to Hegel's philosophy, in that the ideal remains the asymptotic model of the real, but its political use becomes stronger. "It may be urged that every individual man carries, within himself, at least in his adaptation and destination, a purely ideal man. The great problem of his existence is to bring all the incessant changes of his outer life into conformity with the unchanging unity of this ideal. This pure ideal man, which makes itself known more or less clearly in every subject, is represented by the state, which is the objective and, so to speak, canonical form in which the manifold differences of the subjects strive to unite. Now two ways present themselves to the thought, in which the man of time can agree with the man of idea, and there are also two ways in which the state can maintain itself in individuals. One of these ways is when the pure ideal man subdues the empirical man, and the state suppresses the individual, or again when the individual becomes the state, and the man of time is ennobled to the man of idea" (Schiller 1910: letter IV). Even though Fichte keeps believing that the ideal cannot be made real ("the concept of man is an idealistic concept, because man's end qua man is something unachievable," Fichte 1993: 157),[5] his political philosophy, that is, his political use of the ideal, is important because of its epistemological revolution.

Indeed, we may say that the words of Schiller that explicitly refer to Fichte's idealistic philosophy summarize the two kinds of consequences that can logically result from the crisis of the relationship between reality and ideality, that is, the removal of their epistemological dualism, in that:

1. The first possible consequence is that, in Schiller's words, "the pure ideal man subdues the empirical man, and the state suppresses the individual." This is precisely the epistemological structure of any form of totalitarianism: first, we remove the epistemological dualism between reality and ideality; second, we try to get to a unique ontological dimension, which is the ontological dimension of ideality, by trying to make the ideal real; third, in particular, we try to embody the ontological qualities of the ideal, that is, universality, perfection, completeness, and invariability into the real, whose ontological qualities we try to dissolve (see especially the manifestos of the political totalitarianisms we have experienced, which clearly show to be founded on this epistemological structure).[6] Yet, as European history proves, trying to make the ideal real necessarily results in dramatic failure: in particular, a dramatic failure following tragic violence used in an attempt to do what is logically impossible.
2. The second possible consequence is that, in Schiller's words, "the individual becomes the state, and the man of time is ennobled to the man of idea." This is precisely the epistemological structure of any form of anarchism: first, we remove the epistemological dualism between reality and

ideality; second, we try to get to a unique ontological dimension, which is the ontological dimension of reality, by trying to make the real ideal; third, in particular, we try to embody the ontological qualities of the real, that is, particularity, imperfection, incompleteness, and variability into the ideal, whose ontological qualities we try to dissolve. As I shall try to argue, the Western culture, as a reaction to the dramatic failure of any attempt to make the ideal real, seems to increasingly experience forms of anarchism, which does not mean that the ideals are relative (as we have seen, even Kant thinks that the ideals are relative, in that they start from our empirical experience), but that the ideals are dissolved—and, in its most radical form, as we are experiencing in our technological era, anarchism means not only that the superior ontology of ideality is dissolved but also that we are replacing it by technologically creating an ontology that is, outwardly, superior by being omnipresent, omniscient, and omnipotent, but, inwardly, inferior by being totally immanent and, in particular, our totally immanent creation.

If we go back to the modern Prometheus, then we can identify an essential correlation between what happens to the symbol of the relationship between humans and technology and what happens to the Western culture and, in particular, to Western epistemology. In the latter case, ideality is thought of not as reality's best tool to develop its ontology, but as reality's ontological destiny, by trying to move from imperfect incompleteness, variability, and particularity to perfect completeness, invariability, and universality. In the former case, the modern Prometheus, that is, Frankenstein, uses technology to try to move from human ontology (which is imperfect, incomplete, variable, and particular) to divine ontology (which is perfect, complete, invariable, and universal) both by creating living human matter from nonliving matter and by making his creature resistant to diseases and death itself.

Moreover, among the details we have seen in the first chapter, one is crucial: moving from the ancient Prometheus to the modern Prometheus means trying to move, through technology, from two ontological dimensions (which are the human and the divine, as well as the real and the ideal) to a unique ontological dimension (which is the divine, as well as the ideal). More precisely, whereas ancient culture distinguishes the punisher (Zeus, i.e., the superior ontology) from the punished (from Prometheus and humans, i.e., the inferior ontology), modern culture makes no ontological distinction at all. If we ask who punishes, then we should answer: Frankenstein, humans, and Frankenstein's creature. And if we ask who is punished, then we should answer: Frankenstein, humans, and Frankenstein's creature. The epistemological revolution is showed not only by the unique ontological dimension, in that everyone is equally the punisher and the punished, but also by what happens to technology in particular. In ancient culture, technology (i.e., fire) is neither the punisher nor the punished, whereas in modern culture, technology (i.e., Frankenstein's creature) is both the punisher and the punished—the epistemological revolution, that is, moving to a unique ontological dimension, means that the more technology develops, the more indiscernible both the ontological distinction between the human and the divine and the ontological distinction between the human and the technological become. In the former case, the modern Prometheus shows us that, through an ambitious development of technology, humans can both create and prevent diseases and death itself and, therefore, equal the divine. In the latter case, the modern Prometheus shows us that, through an ambitious development of technology, technology can both punish and be punished and, therefore, equal the human—technology can move from being an object to being a subject (and, as we shall see, technology can also move from equaling the human to equaling the divine that humans are trying to equal).

CONTEMPORARY TECHNOLOGY AS A CREATION OF AN IMMANENT DIVINE

Going back to Schiller's words, the second possible consequence that can logically result from the crisis of the relationship between reality and ideality, that is, the removal of their epistemological dualism, is the following: "the individual becomes the state, and the man of time is ennobled to the man of idea." We have seen that this is precisely the epistemological structure of any form of anarchism, which is the crucial issue through which I am trying to read our technological era.

Let us consider the etymological meaning of the word "anarchism," starting from the etymological meaning of the word "anarchy," the former being the radicalization of the latter. "Anarchy" means absence (ἀν) of something that rules (ἄρχω)—"anarchism" means radical "rulerlessness," that is, radical absence of what the Western culture has, for millennia, entrusted to a superior ontology in general and to ideality in particular, sometimes from a chronological perspective (by making reference to a ruler as a beginning) and sometimes from a logical perspective (by making reference to a ruler as a cause). From a chronological perspective, ideality is the beginning, in that it is the matrix of reality, that is, the definite article (universal, identifying what should be) that is the matrix of the indefinite articles (particular, identifying what is). From a logical perspective, ideality is the cause, in that it is the condition of possibility for reality to do something essential for its development: first, reality can identify its present identity by comparing it with ideality; second, reality can identify the future development of its identity by comparing its imperfection with ideality's perfection and, consequently, by asymptotically perfecting the former through its comparison with the latter. In any case, their difference is essential: reality needs ideality to be its perfect counterpart.

Anarchism radically abolishes ideality as reality's perfect counterpart. We may say that this is precisely what is increasingly happening in the European culture in particular and in the Western culture in general after having experienced the first tragic failure of totalitarianism as the first tragic failure of an attempt to make the ideal real.[7] More precisely, as we shall see in the following pages, what increasingly happens is that ideality starts to be thought of as dramatically dangerous and, in particular, as something that must be removed. This is the beginning of anarchism—European culture, in particular, and the Western culture, in general, having invented and successfully used ideality for millennia as reality's best tool to develop its ontology, first, denaturalizes it and, second, abolishes it.

Speaking of this kind of anarchism does not mean speaking of relativism. Indeed, their epistemological structures are different. Let us go back to the simple example of the craftsman, and let us imagine that we are craftsmen, too. We want to answer the following question: what should we do to make a good couch? We can answer in three different ways:

1. There is one possible acceptable way of making a good couch, that is, ours (A), and not the craftsman's (B). Thus, we ask the craftsman to follow our way, too.
2. There are several possible acceptable ways of making a good couch, that is, ours (A), the craftsman's (B), and other peoples' (C, etc.). Thus, we follow our way and do not ask anyone to follow our way, too.
3. There is no acceptable way of making a good couch, that is, there is no ideal couch, whether identical for us, the craftsman, and other people or different for us, the craftsman, and other people. Thus, we make our couch by improvising and do not ask anyone anything.

In the first case, we exercise a form of totalitarianism. In the second case, we exercise a form of relativism. In the third case, we exercise a form of anarchism.

Relativism and anarchism share something important: both as relativists and as anarchists, we do not ask anyone to follow our way. But their difference is more important. As relativists, we believe that ideality keeps guiding us: our ideal couch (A) guides us when we make our couch, the craftsman's ideal couch (B) guides him when he makes his couch, and other peoples' ideal couches (C, etc.) guide them when they make their couches. If we choose a weak form of relativism, then we think that it is useful to talk to others to try to compare our ideal couch (A) with their ideal couches (B, C, etc.) since we think that, by comparing them, we may perfect our ideal couch (A) and, consequently, the couches we are going to make in the future. If we choose a strong form of relativism, then we think that it is not useful to talk to others to try to compare our ideal couch (A) with their ideal couches (B, C, etc.) since we think that they are so different that they are incomparable. On the contrary, as anarchists, we think that ideality does not keep guiding us at all: the epistemological solipsism of the strong form of relativism is radicalized, in that we think that there are no ideal couches to compare at all. Reality, together with its ontological qualities, that is, particularity, imperfection, incompleteness, and variability, is all that exists. As such, we will improvise our contingent couch, the craftsman will improvise his contingent couch, and other people will improvise their contingent couches. Moreover, we will neglect the craftsman's couch and other people's couches: if we think that contingent reality is all that exists, then we have no reason at all to think that comparisons may be useful, since everyone is guided uniquely by their own contingency. Thus, the negative consequences of anarchism are worse than the negative consequences of strong relativism. The latter allows us to make at least one useful comparison: the comparison between our own couch and our own ideal couch, which can help us perfect our own couch. On the contrary, the former does not allow us to make useful comparisons at all: anarchists' couches are extremely solipsistic, contingent, and accidental—anarchists' actions, in general, are extremely solipsistic, contingent, and accidental. As such, they easily stagnate and, moreover, hardly improve: anarchists' future couches are going to be as solipsistic, contingent, and accidental as their present couches.

As I shall try to argue, the kind of anarchism that characterizes our technological era is special: it is the most radical form of anarchism we have ever experienced. However, Paul K. Feyerabend's description of the epistemological anarchist in his *Against Method. Outline of an Anarchistic Theory of Knowledge* can give us important insights: "the one thing he opposes positively and absolutely are universal standards, universal laws, universal ideas such as 'Truth,' 'Reason,' 'Justice,' 'Love,' and the behaviour they bring along" (Feyerabend 1975: 189) since "there is only *one* principle that can be defended under *all* circumstances and in *all* stages of human development. It is the principle: anything goes" (Feyerabend 1975: 28). He adds in his *Farewell to Reason*: "'Truth', written 'in capital letters,' is an orphan in this world, without power and influence" (Feyerabend 1987: 102). Feyerabend describes in philosophical terms and, specifically, in epistemological terms a cultural phenomenon in general, which is somehow anticipated by avant-garde art, which seems to be able to express the practical consequences that the theoretical crisis of ideality can have. As Feyerabend himself writes: "like the Dadaist, [. . .] he [the epistemological anarchist] 'not only has no programme, [he is] against all programmes'" (Feyerabend 1975: 189),[8] and he "becomes capable of stepping outside the most fundamental categories and convictions, including those which allegedly make him human" (Feyerabend 1975: 189). If we make reference to the *Dada Manifesto*, then we read: "I am against systems, the most acceptable system is the one of not having any system, on principle" (Tzara 2001: 299). Yet, "I am also against principles" (Tzara 2001: 300). In any case, "There is no final Truth" (Tzara 2001: 298) and there is no ideal: "If I shout: IDEAL, IDEAL, IDEAL, KNOWLEDGE, KNOWLEDGE, KNOWLEDGE, BOOMBOOM, BOOMBOOM, BOOMBOOM, I have put down rather exactly the progress, the laws, morality, and all the other lovely qualities that various very intelligent people have discussed in so many books, just in order to say finally that each man has danced anyway according to his own personal boomboom, and that he is right in his boomboom" (Tzara 2001: 302). And the most important practical consequence is the following: "So Dada was born for a desire for independence, of a distrust of the community. Those who belong to us keep their freedom" (Tzara 2001: 302). Of course, an epistemological attitude (a way of thinking) underpins an ethical attitude (a way of acting): if there is no ideal universality sharable by real particularities, then any real particularity is "independen[t]" and "free" (yet, we may say that it is "independen[t]" and "free" not by choosing a particular option from several others, but by choosing the only option available, that is, the solipsistic, contingent, and accidental option that coincides with its equally solipsistic, contingent, and accidental situation).

Let us consider at least one more philosophical description of anarchism, offered by Richard Rorty in his *Consequences of Pragmatism. Essays, 1972–1980*. The philosopher he thinks of deserts the ideal of doing something truly meaningful (being a philosopher does not mean "having actually solved some of the 'problems of philosophy'," Rorty 1982: 219), deserts the ideal of doing something more meaningful than playing a kind of rhetorical game (being a philosopher means being "able to construct as good an argument as *can* be constructed for *any* view, no matter how wrong-headed," Rorty 1982: 219), and deserts the ideal of founding their thoughts and actions on something that distinguishes what is good from what is wrong (being a philosopher means being a kind of barrister who is able to "construct, or criticize, any argument," no matter what the "argument" is, Rorty 1982: 219).

Indeed, Rorty's philosopher is not surreal at all: their description is, outwardly, somewhat hyperbolic but, inwardly, a kind of portrait of several real contemporary philosophers who seem to embody epistemological anarchism into their professions. This happens, for instance, when a philosopher identifies the profession with trying to find an extremely limited issue on which to try to say something (i.e., anything) in an original way, no matter what the issue is and, moreover, no matter what its scenario is. Outwardly, the reason for this attitude is ambition, even narcissism: the more originality the philosopher can show to have (in the easiest way, i.e., on an extremely limited issue and uniquely rhetorically), the more distinguished the position is supposed to be. Inwardly, there is another reason for this attitude to add: epistemological anarchism. The philosopher focuses on rhetorical originality on an extremely limited issue because there is nothing left to identify the meaning of the profession. More precisely, there is no ideal guiding the work—there is no ideal philosophical scenario guiding the work. It is no coincidence that a degree in philosophy leads with increasing frequency to a communication job analogous to that of a barrister, who is able, and available, to "construct, or criticize, any argument," no matter what the "argument" is (in Feyerabend's words, no matter what the "programme" is since "anything goes." In Tzara's words, no matter what the "system" is and the "principles" are since "there is no final Truth"). Thus, the philosopher, analogously to the barrister, ends up claiming any "argument" at all, that is, an extremely solipsistic, contingent, and accidental "argument," which makes their profession equally solipsistic, contingent, and accidental: if there is no ideal philosophical scenario as a guide to argue for, then the philosopher can claim "argument" A today and "argument" non-A tomorrow since, for instance, today's contingent job market advantages A and tomorrow's contingent job market advantages non-A. Outwardly, this seems to be "independence" and "freedom." Inwardly, this is a severe kind of dependence on what may be the worst kind of guide, i.e., extreme contingency (as well as a severe kind of mercenarism).

If we go back to a possible contemporary Prometheus, then we can identify an essential correlation between what happens to the symbol of the relationship between humans and technology and what happens to the Western culture and, in particular, to Western epistemology. In the latter case, ideality, as well as any kind of superior ontology, is increasingly abolished by reality, which makes itself increasingly absolute by causing its ontological qualities to occupy all that exists: particularity and contingency increasingly make all that exists into a unique ontological dimension in which everything is equal because everything is equally particular and contingent. In the former case, a possible contemporary Prometheus, that is, for instance, the trinity of the creators, uses technology to try, outwardly, to do something analogous to what the modern Prometheus does, by trying to move from human ontology to divine ontology ("Now we're all God": "We will become all-seeing," "all-knowing," and "perfect or near to it" by "solv[ing] any problem," "cur[ing] any disease," and "end[ing] hunger"). But, inwardly, there is a crucial difference between what Frankenstein does and what the trinity of the creators does. The former tries to embody the divine into the human (as well as ideality into reality) both by creating living human matter from nonliving matter and by making his creature resistant to diseases and death itself. The latter tries to embody a special kind of the divine into a special kind of human, that is, a divine that is, as it were, humanly made into a human that is, as it were, divinely maker—again, the ontological distinctions dissolve into a unique ontological dimension in which the human makes the divine as a technology that can be omnipresent, omniscient, and omnipotent by being totally immanent, that is, that can be universal by being totally particular.

More precisely, a possible contemporary Prometheus shows us the following:

1. A trinity of creators both human (by being human beings) and divine (by being a trinity of creators who obtain, through technology, the divine prerogatives of omnipresence, omniscience, and omnipotence), as well as technological (by being human beings who end up being cyborgs becoming increasingly indiscernible from the technological devices of "the Circle").
2. A technology (the technological devices of "the Circle") both technological (by being a technological creation) and divine (by being omnipresent, omniscient, and omnipotent), as well as human (by being a human creation becoming increasingly indiscernible from human beings).
3. Moreover, a "God" both divine (by being omnipresent, omniscient, and omnipotent) and technological (by being a technological creation), as well as human (by being a human creation becoming increasingly indiscernible from human beings).

It is no coincidence that technology and "God" can be described by identical words: the latter ends up being the former, that is, the most immanent thing humans have ever made (in Eggers' words, a thing made by humans who, "with ruthless capitalistic ambition," "professionalized our idealism, monetized our utopia").

If we move from the kind of technology symbolized by a possible contemporary Prometheus to our technological era, then we realize that our notion of technology can be understood as the most radical form of anarchism we have ever experienced. Before taking into account cases in point (in the third chapter), let us consider at least an example through which we can start to reflect upon our notion of technology. Let us imagine that we have a symptom we have never had before. What do we do? The answer is that we take our smartphone, google our symptom, and self-diagnose. Then, the less hypochondriacal among us says: "Okay, it's nothing serious. I won't go to the doctor." And the more hypochondriacal among us says: "Okay, it's nothing serious. But I'll go to the doctor anyway, because Google says that my symptom may be serious in exceptional cases." And when the more hypochondriacal among us goes to the doctor, the doctor says: "Your symptom isn't serious. Don't worry about it, just forget about it." And we say: "But Google says that my symptom may be serious in exceptional cases. Can you make me take some tests?." And the doctor, quite annoyed, says: "Absolutely not. If you want to take some tests just because Google says you should, then you shouldn't be using public health care services." Then, we go home with our tail between our legs with no idea of what to do since we cannot afford private health care to take the tests. (Then, the doctor secretly googles our symptom.) Over the next days, we see several automatic advertisements about our symptom on our smartphone and get scared. As a result, we rack up debt just to be able to afford private health care and take the tests. Finally, the tests come back negative: our symptom, which has vanished in the meantime, was nothing at all. This example may seem comical, but it has a dramatic side, and it is becoming increasingly common nowadays.

If we compare this example with the results of our reflections upon the meaning of the correlation between a possible contemporary Prometheus and Western epistemology, then we can argue the following:

1. As for humans, we should consider at least ourselves and the doctor (in addition to the creators of Google, of course. But there may be an easier correlation between them and the trinity of the creators analyzed). Epistemological anarchism can help us understand both our behavior and the doctor's behavior. As for our behavior, we use technology (Google) to remove the role of the expert (of the doctor) as the mediator between our ignorance and the truth. Thus, by removing the role of the expert, we think that we can get to the truth directly. More precisely, we think that technology can give us the expertise of the expert—we may even say that we think that technology can make us omniscient. Thus, we are radical epistemological anarchists in that our solipsism is so strong that we even end up self-diagnosing without being doctors. If, contingently and accidentally, we are not hypochondriacs, then our solipsistic solution is totally satisfactory and we do not go to the doctor. On the contrary, if, contingently and accidentally, we are hypochondriacs, then our solipsistic solution is not totally satisfactory and we go to the doctor. However, in both cases, our notion of truth experiences a severe crisis. Even though we use technology to become omniscient, this kind of omniscience is totally immanent (i.e., particular and, therefore, contingent and accidental), which means, for instance, that Google's algorithm may assert, today, that our symptom is not serious and, tomorrow, that our symptom is serious. In the former case, the crisis of our notion of truth also shows up when we believe that our expertise is equal to the doctor's expertise. In the latter case, the crisis of our notion of truth also shows up when we continuously confuse, on the one hand, anxiety and truth and, on the

other hand, our expertise, the doctor's expertise, and Google's expertise. As for the doctor's behavior, the risk is that they use technology to remove their role of expert as well: they use technologies (sometimes also Google) that risk to make them into a kind of anarchic bureaucrat (as we shall see in the third chapter).

2. As for technology, it may be sufficient to say that the kind of technology we are considering (Google) is even the verb that means the kind of action we do ("to google"). Epistemological anarchism can help us understand the phenomenon described, in that its total immanence makes its omnipresence (it is everywhere we are), omniscience (it knows everything we want to know), and omnipotence (it has power over everything we do) ambiguous. In particular, its omniscience is founded on an algorithm that is, in any case, particular and, therefore, its knowledge is, in any case, contingent and accidental (again, it may assert, today, that our symptom is not serious and, tomorrow, that our symptom is serious). Google clearly exemplifies the special form of epistemological anarchism we are reflecting upon. On the one hand, it is a kind of omnipresent, omniscient, and omnipotent God by being, on the other hand, a totally immanent human creation. In particular, on the one hand, it knows everything by getting, on the other hand, to a contingent and accidental truth that is particular and, therefore, imperfect, incomplete, and variable—we may summarize by saying that this special form of epistemological anarchism is the most radical form of anarchism we have ever experienced precisely because what we are actually doing is creating reality as a unique ontological dimension by creating a special kind of divine that embodies, outwardly, the typical ontological qualities of ideality and, inwardly, the typical ontological qualities of reality.
3. As for the divine, again, it is no coincidence that we can use identical words to describe it and technology. But we should add something we shall focus on in the third chapter: the total immanence of this special kind of divine has crucial consequences on our behaviors and lives. As soon as we have the symptom, we take our smartphone to google it and self-diagnose. And, as soon as we leave the doctor, Google sends us several automatic advertisements about our symptom on our smartphone. We may say that this special kind of divine is, fairly literally, a pocket divine (which may be also a good image for representing its total immanence). The first consequence of being a pocket divine is that we lose the most important possibility of infinite progressive improvement, which is something that, as we have seen, only an asymptotic model can give us. If we refer to an ideal model, as well as to a transcendent divine, then we are required to asymptotically improve our abilities by continuously training them. In our case, without a smartphone in our

pocket, we are required to continuously train at least the following: first, our attention (so that we can understand, for instance, that this symptom shows up when we eat a specific food); second, our memory (so that we can remember it for the future); third, our ability to control our anxiety (so that we can prevent our anxiety from interfering with our memory. And, in the worst case, so that we can wait to talk to the doctor without panicking); fourth, our critical thinking (so that we can believe that if any diagnosis can be asymptotically true, then we should not make any diagnosis absolute and, therefore, we should talk to another doctor, at least if the first doctor's diagnosis is critical). On the contrary, if we refer to a real smartphone, as well as to an immanent divine, then we end up stopping to continuously train our abilities. I am not at all arguing that we should continuously train all of our abilities (I may be wrong, but I am pleased to stop to continuously train my memory using one of the first technologies we created, i.e., writing). On the contrary, I am arguing that we should consciously decide what abilities we should continuously train since we believe that they are necessary to be who we want to be. For instance, if I consciously decide what abilities I should keep training, then my answer is that I should keep training my critical thinking. However, it is not as simple as it may seem. I have a smartphone, and I am a hypochondriac, and, as such, I am constantly risking various things: first, to shorten my attention span (because when a symptom shows up, I can immediately google it); second, to weaken my memory (because I can immediately google it again and again); third, to weaken my ability to control my anxiety (because I can immediately find information without waiting to talk to the doctor); fourth, and most severely, to weaken my critical thinking (because if I have no strong attention span, memory, and ability to control my anxiety, then I can prevent myself from panicking only by thinking that any information I can immediately find is absolutely true).

It is no coincidence that I used the word "immediately" four times in the last paragraph. Its etymology is most clear: it means that I continuously risk to use technology (Google, in this case) to remove, as we have seen, the role of the expert (the doctor, in this case) as the "mediator." And, of course, if there is the "mediator," then there is no "immediacy," and I cannot do anything "immediately." Again, "immediacy" means anarchism: there is "immediacy" if there is a unique ontological dimension in which the ontological distinctions (for instance, between my ability to self-diagnose through Google and the doctor's ability to diagnose) dissolve—"immediacy" means anarchism because it means that everything is equal by being equally particular and contingent (but it may be severely wrong).

NOTES

1. In the following passage:

> the mind makes particular ideas received from particular things become general; which it does by considering them as they are in the mind—mental appearances—separate from all other existences, and from the circumstances of real existence, such as time, place, and so on. This procedure is called abstraction. In it, an idea taken from a particular thing becomes a general representative of all of the same kind, and its name becomes a general name that is applicable to any existing thing that fits that abstract idea. Such precise naked appearances in the mind, without considering how or from where or in company with what others it acquired them, the understanding stores away for use as standards: it will classify real things into sorts on the basis of their agreement with these patterns or standards. The abstract ideas have names commonly attached to them, so that they also serve as patterns for applying words, labels, to the things that they enable us to sort. Thus you observe the same colour today in chalk or snow that you yesterday saw in milk; your mind considers that appearance alone, makes it a representative of all of that kind, and gives it the name "whiteness"; and by that sound you signify the same quality, wherever it is imagined or met with. This is how universals, whether ideas or words, are made. (Locke 1690: II, 11, 9)

2. In the following passage:

> To understand this more clearly, let us trace our notions and names from their beginning in infancy, and see how they develop from there. It's perfectly obvious that the ideas of the persons that children encounter are, like the persons themselves, only particular. The ideas of the nurse and the mother are well formed in an infant's mind. They represent only those individuals, and the only words that the infant has for the individuals are, in effect, proper names, like "Nurse" and "Mamma." As they get older and meet more people, infants notice that many other things in the world resemble—in shape and in other ways—their father and mother and other people they have been used to; and they form an idea that applies equally to all those many particular people, associating this idea with the name "man." That is how they come to have a general name and a general idea. In doing this, they don't make anything new, but only leave out of the complex ideas they had of Nurse and Mamma, Peter and James, Mary and Jane, whatever is unique to each, and retain only what is common to them all. (Locke 1690: III, 3, 7)

3. As we have seen. See at least Arthur (2011) and Latour (1987).

4. In particular, being the most defining feature distinguishing the Western culture from the Chinese culture. See at least Cheng (2004), Ghilardi (2014), Jullien (1995, 1998, 2000, 2006, 2008 and 2009), Lloyd and Sivin (2002), Longxi (1992), Pasqualotto (2003). See also Chiodo (2018).

5. Thus, "the scholar has reason to be the humblest person of all: since the goal which is set for him must always remain very distant, and since he has to achieve a very lofty ideal—one from which he normally remains very distant" (Fichte 1993: 169–70).

6. See especially Gentile's words: "Idealism [is] understood as faith in the advent of an ideal reality, as a manner of conceiving life not as fixed within the limits of existing fact, but as incessant progress and transformation toward the level of a higher

law which controls men with the very force of the idea" (see Gentile 1929, available in Project Gutenberg: https://www.gutenberg.org/files/14058/14058-h/14058-h.htm #THE_PHILOSOPHIC_BASIS_OF_FASCISM). See also Marx and Engels 1991.

7. Interestingly enough, totalitarianism's tragic failure caused not only the attack on ideality and, in particular, on its being made absolute but also the attack on technology as a privileged way of establishing capitalistic forms of totalitarianism (see especially the Frankfurt School: Horkheimer and Adorno 2002; Marcuse 1991).

8. Feyerabend refers to Richter (1965).

Chapter 3

Reading Our Technological Era

A FIRST CASE IN POINT: ALGORITHM AND HEALTH

The second chapter ended with an imagined example (even though it may be applicable to several situations). Let us start the third chapter with a real-life case, that is, one from my own experience.

About three years ago, a member of my family was told by the staff of a good public hospital in northern Italy that she had the chance to take a genetic test since there had been cases of a specific type of cancer in her family. The hospital used a protocol to distinguish those who could take it from those who could not take it. The first step of the protocol was filling in a questionnaire about her family's health history and, specifically, ages and causes of death of her older relatives. I helped her fill in the questionnaire: several pieces of information were necessarily imprecise, sometimes because they were not available at all (for instance, if a family member died several decades ago without an official diagnosis) and sometimes because we could not remember them (for instance, if there was an official diagnosis we could not remember). The second step of the protocol was the analysis of the questionnaire we brought back to the hospital. Finally, we received an official letter from the hospital in which it was literally written that, according to the "predictive algorithm" that processed the data from the questionnaire, she was not eligible to take the genetic test, since her risk to be genetically predisposed to suffer from that specific type of cancer was not high enough. Of course, the decision of the "predictive algorithm" was meaningful not only for her life and future but also for her relatives and, in particular, for her descendants (myself included): indeed, the genetic test would have returned meaningful information for all of us.

There is a clear analogy between this case and the imagined example described in the second chapter: in both cases, a sophisticated technology has great power over our lives and futures (outwardly, by helping us make decisions and, inwardly, by making decisions for us).

I think that it is useful to start from what really happened when we received the letter from the hospital. Thus, let us move from the imagination to the fact. As for the member of my family to whom the letter was addressed, she did not have a clear idea of what an algorithm was, like most people her age. However, according to the situations in which she had heard about algorithms, she had the feeling that an algorithm was something we could trust. More precisely, she had the feeling that an algorithm was something more reliable than a human being. As for me, I knew somewhat more about what an algorithm was (at least because I work at the Politecnico di Milano). For instance, I knew that an algorithm works on past data, which means that if an algorithm says that my risk to suffer from a disease is low, but I dangerously change my lifestyle in the future, then the prediction of the algorithm is likely to be false, that is, I am likely to suffer from the disease in the future. However, I realized that I felt reassured when I read that, according to the "predictive algorithm," the risk of the member of my family to be genetically predisposed to suffer from that specific type of cancer was low. I even realized that my first reaction was the following thought: "I can avoid checking my health annually, and feeling anxious annually waiting for the results of the tests." Moreover, both the member of my family and I thought that an algorithm could be more reliable than a human being and, in particular, than several human beings we knew (doctors included), since the former can be absolutely stable and the latter can be absolutely unstable (for instance, a human being can make a mistake for the most idiosyncratic emotional reasons).

In addition to what really happened when we received the letter from the hospital, there is at least one other meaningful detail to consider, which has to do with what we have reflected upon in the second chapter: the dissolution of the role of the expert as the mediator—again, the dissolution of the mediator as anarchism. What really happened to us was that no human being talked to us: we received the letter without any explanation from any human being. The reason why this detail is most meaningful is that it clearly shows us not only the kind of epistemological anarchism we are increasingly experiencing but also the kind of ethical anarchism, as a consequence of epistemological anarchism, we are increasingly experiencing. The fact that a letter, and not a doctor, says that an algorithm and, in particular, a "predictive algorithm," and not a doctor, decides that the member of my family is not eligible to take the genetic test is meaningful for the following epistemological and ethical reasons.

First, from an epistemological perspective, we are facing a progressive epistemological discharging. More precisely, as we shall see in the fourth chapter, we can call this phenomenon epistemological externalization—epistemological externalization from the human to the technological as an immanent divine. In our case, the following details are meaningful:

1. The use of the protocol. Contemporary medicine is massively, and increasingly, characterized by the use of protocols. In general, the use of a protocol means that a doctor is not alone when they decide what to do. On the contrary, they have something more reassuring than their own wisdom, and even than their colleagues' wisdom: the protocol, which is the set of rules they have to follow. The set of rules is written and shared, saying that, if the patient is in given situation A, then they must do given thing B, etc. Setting, writing, and sharing rules mean moving from something internal to something external, that is, from their decision to a decision we can call, especially in our case, technological. In our case, when they use the "predictive algorithm" to decide if the member of my family is eligible to take the genetic test, they externalize their decision in a meaningful way: if something goes wrong, then they are not definitely responsible for the possible epistemological mistake, as well as for its possible ethical consequences. In particular, they are not definitely prosecutable by law. Of course, I am not at all saying that contemporary medicine should not use protocols, which can be remarkably useful in several ways (see at least Field and Lohr 1990; Mace 2017). For instance, they can guarantee that, when a doctor is exhausted after having worked for several hours, they can be helped in avoiding mistakes caused by a lapse in concentration: indeed, they can follow the protocol without the risk to overthink. On the contrary, I am saying that contemporary medicine should not end up making protocols absolute but should keep making them work, in any case, together with the doctors' minds (which keeps happening, but progressively less. And which was crucial, for instance, in the case of the first Italian diagnosis of coronavirus: a doctor could diagnose it precisely by not following the protocol).[1] As we shall see in the fourth chapter, a radical epistemological externalization would dangerously mean, first, that we would end up neglecting the precious epistemological resources of the human mind, not shared by an algorithm, and, second, that we would end up atrophying its precious epistemological resources (which have to do, for instance, with the fact that a doctor, and not an algorithm, may happen to understand, somehow intuitively, that something is wrong with their patient from the look on their patient's face).
2. The use of the questionnaire. It is the tool used to obtain the data that the algorithm processes. It is crucial from an epistemological perspective, in

that it is through the data from the questionnaire that the particular (i.e., the individual pieces of information) can be compared with the universal (i.e., the algorithm and, in particular, its rules for processing the data) and, therefore, identified. But, as we have seen, the questionnaire itself is critical: first, sometimes the individual pieces of information are not available at all; second, sometimes the individual pieces of information cannot be remembered. Moreover, at least in our case, the questions of the questionnaire were focused on the past (i.e., older relatives) but not on the present (i.e., relatives belonging to the same generation, as well as descendants). However, the eligibility of the member of my family to take the genetic test was decided by the "predictive algorithm" that processed the data from the questionnaire. We may think that there is a kind of paradoxical epistemological advantage for both the doctors as human beings and the hospital as an institution, in that, if something goes wrong (for instance, if what an algorithm decides to do in a given case turns out to be wrong), then both the doctors and the hospital can say that the epistemological mistake was made by the patient who filled in the questionnaire, and not by themselves, who are not responsible for any epistemological mistake, as well as not prosecutable by law.

3. The use of the algorithm and, in particular, of the "predictive algorithm." From an epistemological perspective, we have to consider at least three kinds of critical issues. First, as we have seen (and as we shall see especially in the fourth chapter), making the use of the algorithm absolute may result in neglecting precious epistemological resources that the human mind, and not the algorithm, has. Second, as several authors highlight (see at least Domingos 2015; Noble 2018; O'Neil 2016; Rees 2018; Ross 2016, among others), algorithms are anchored to the past, that is, the past data through which they have been trained. Thus, their prediction of the future may be seriously limited, even though, paradoxically enough, it is actually their most important aim (on the contrary, we may ask ourselves the following provocative question: are we actually sure that the human mind is equally anchored to the past?). Third, at least in our case, the use of the word "predictive" may be dangerous from an epistemological perspective, in that it risks to make a promise that it is unable to keep. Thus, most people risk to believe that something that may be false is absolutely true (and sometimes the ethical consequences of this epistemological mistake are dramatic).

Again, epistemological anarchism can help us understand the phenomenon described, that is, epistemological externalization. Indeed, a fairly easy way of getting to the radical "rulerlessness" meant by "anarchism" is, first, to externalize what rules by shifting it from our minds, where it is an ideal, to the outside world, where it is an algorithm, and, second, to make it totally immanent. The ultimate result is that we get to the kind of Dadaist "freedom" desired by all anarchists. Even though, as we have seen, it is a dangerous kind of "freedom," it can surely provide us with a freeing delegation that seems to be one of contemporary technology's essential aims. Indeed, we are using technology not only to remove the role of the expert as the mediator but also to free ourselves from epistemological (and ethical) responsibility after having removed the mediator: if we are the patients and something goes wrong, then we can say, at least, that it is the algorithm's fault, and not ours. As such, the phenomenon described can be read as the most radical form of anarchism we have ever experienced precisely because it is not only a matter of removing a ruler that can make us realize that we make mistakes—it is a matter of trying to remove the very fact that we make mistakes at all.

Second, from an ethical perspective, we are facing a progressive ethical discharging, which we can call ethical externalization from the human to the technological as an immanent divine. In our case, the use of the protocol, the use of the questionnaire, and the use of the algorithm can be read from an ethical perspective as follows:

1. As for the use of the protocol, if I am your doctor and you are my patient, then I can avoid feeling both extremely under pressure when I make the diagnosis and extremely guilty when something goes wrong.
2. As for the use of the questionnaire, I can add that, in any case, you are guiltier than me since you filled in the questionnaire.
3. As for the use of the algorithm, I can even add that, in any case, no human being is actually guilty, since the algorithm processed the data (in particular, I can even add that, statistically, algorithms reduce the incidence of mistakes. And, if you are the exception, for instance, because of something that is not processable, then, again, no human being is actually guilty).

The ultimate result is crucial: saying that no human being is actually guilty (which happens more and more frequently) means experiencing a severe crisis of the very notion of individual responsibility—from an ethical perspective, the most radical form of anarchism we are experiencing means that we are trying to use technology to free ourselves from individual responsibility, that is, one of the most typically human prerogatives, at least in the Western culture, from Aeschylus to Shakespeare to Pirandello (as we shall see in the following chapters).

It is no coincidence that our technological era is characterized by a special kind of difficulty in identifying actual individuals actually responsible for what happens. In our case, both the doctors as individual human beings and the hospital as an institution established by individual human beings are difficult to identify as responsible if something goes wrong. Indeed, from a legal perspective but also, and above all, from an ethical perspective, if the present decision of the "predictive algorithm" turns out to be wrong in the future, then it is difficult to identify the actual individuals actually responsible. Even though we can say that, legally, the chief medical officer of the hospital may be identified as responsible, we cannot equally say that, ethically, they may not only be identified as responsible but also feel responsible.

More precisely, the "predictive algorithm," in particular, and technology, in general, seem to be used more and more frequently to obtain a special kind of thing that can prevent individual human beings from being legally and, moreover, ethically identified as responsible and, in particular, as individually responsible: a special kind of bureaucracy that, again, coincides with a technological entity as an immanent divine. In our case, we can argue that the doctors as individual human beings, by freeing themselves especially from individual ethical responsibility through the "predictive algorithm," seem to end up being a special kind of anarchic bureaucrats: bureaucrats, in that protocols win out over human decisions, and anarchic, in that, again, it is a matter of trying to remove the very fact that mistakes can be made by human decisions at all—anarchic, in that technological criteria and responsibility win out over human criteria and responsibility and, therefore, humans can stop to compare their real criteria and responsibility with ideal criteria and responsibility: they can be, again, "rulerless."

Interestingly enough, bureaucracy was introduced to a different end, that is, as a way to increase rationality and fairness by moving the management of power from one individual human being to several individual human beings: indeed, if the management of power is up to several individuals, then impersonality and impartiality can more easily increase. Thus, rationality and fairness can more easily increase, too (see at least Weber 2013). But the special kind of technological bureaucracy we are reflecting upon is different: by going one step further, that is, by moving the management of power from several individuals to algorithms, it risks to decrease rationality and fairness. More precisely, it risks to neutralize the possibility of supervising rationality and fairness, since algorithms may be black boxes, as they are frequently defined:[2] the reasons why a given input results in a given output are more opaque in the case of algorithms (which we cannot ask) than in the case of individuals (whom we can ask). Again, the kind of technological entity and, in particular, the kind of technological bureaucracy we are reflecting upon may be thought of as an immanent divine, in that it seems to obtain another essential divine prerogative—it seems to obtain inscrutability. And inscrutability is increasing: the more technological bureaucracy increases, the opaquer the reasons why a given input results in a given output become.

Technology's bureaucratic inscrutability may seem to be more a form of totalitarianism than a form of anarchism, in that its opacity may be used by a minority to overrule a majority. Indeed, even though algorithms may be black boxes, their inscrutability is not democratic, as it were (see at least Noble 2018; O'Neil 2016). In our case, the engineers who decide which past data to use to train the algorithm surely know more than the patients, and even than the doctors. In a dystopian scenario, we may imagine that a pharmaceutical company may pay a technology company to train an algorithm through past data in a way that may result in a cost-effective output. In Eggers' dystopian scenario, the reference to totalitarianism is clearest: "it's a totalitarian nightmare," in which "Infocommunism [. . .] paired with ruthless capitalistic ambition" and "professionalized our idealism, monetized our utopia," and in which "90 percent of the world's searches go through the Circle. Without competitors, this will increase. Soon it'll be nearly 100 percent. Now, you and I both know that if you can control the flow of information, you can control everything." Thus, "everyone will be tracked, cradle to grave, with no possibility of escape."

I think that it is necessary to stress two issues:

1. It is surely true that, once the classical epistemological relationship between ideality and reality has started to be revolutionized and, in particular, once ideality has started to experience a severe crisis, we continuously risk to fall into both kinds of extremism. Even though, both chronologically and logically, totalitarianism comes before anarchism, we may risk to fall again into the former after having experienced the latter, since we may risk to think that what rebalances one form of extremism is its opposite form of extremism. Unfortunately, what is happening in European politics is translating this theoretical phenomenon into a real phenomenon (in addition to what is happening globally, facing the coronavirus emergency. See Harari 2020).
2. In any case, I believe that there is an essential difference between the classical forms of totalitarianism we experienced in the twentieth century and the form of totalitarianism that may result from the bureaucratic inscrutability of technology. In the former case, there is an explicit ideal and, in particular, an explicit ideology, which is thought to be something to make real. In the latter case, there is no explicit ideal at all. There is only what we have seen in the second chapter: solipsistic, contingent, and accidental situations. For instance, the solipsistic, contingent, and accidental interest of a pharmaceutical company that wants to make money. For instance, the solipsistic, contingent, and accidental interest of an extremely ambitious engineer who wants to have fun and be the first to make an equally ambitious technology. In any case, immanence (i.e.,

> particular reality) seems to replace transcendence (i.e., universal ideality). We retain a kind of divine: a technological entity that is not only omnipresent, omniscient, and omnipotent but also inscrutable through its special kind of bureaucracy. However, at least in most cases, we do not retain an actual transcendence, that is, an actual universal ideality: the technological entity we are increasingly creating seems to be our way of trying to become "rulerless." Thus, we not only self-diagnose but also self-do, as it were, almost everything: we increase solipsistic, contingent, and accidental idiosyncrasies that end up being the most radical form of narcissism we have ever experienced. Indeed, if there is radical anarchism, then it is possible that there is also radical narcissism. In our case, if we are "rulerless," in that we use technology to self-diagnose, then it is possible that we are also narcissistic, in that we think not only that we do not necessarily need the expert but also that we are sufficiently expert ourselves.

The theoretical structure of narcissism correlates with the theoretical structure of anarchism for another meaningful reason: from an epistemological perspective, narcissistic solipsism is founded on the dissolution of a ruler that, if shared by several individuals, then makes them into an actual society. The ruler is typically a superior ontological dimension we may call ideality in general. If we share a ruler (an ideal, a god, etc.), then we have something that works as a common denominator. I have a reason to compare my behavior with yours, and you have a reason to compare your behavior with mine. Moreover, we have a tool to judge our behaviors both by comparing mine with yours, since they share the same ruler, and by comparing them with the ruler we share—which means, as we have seen, that we have a powerful tool to make our behaviors improve. But, if the ruler dissolves, then we have no reason to compare anything, no tool to judge anything and no tool to improve anything, since our behaviors are incomparable and, therefore, incommensurable—what can make them comparable and, therefore, commensurable, is a common denominator, that is, a ruler, a superior ontological dimension that can be a common matrix, as we have seen, by being a universal ideality (for instance, a rule of behavior), and not a particular reality (for instance, a behavior).

Indeed, incommensurability is a typical feature of anarchism (see at least Feyerabend 1975, 1987; Rorty 1982, 1989). Thus, we may ask why our technological era is characterized by an obsession with sharing, especially through social networks. The words "social networks" seem to oppose the anarchic solipsism, narcissism, and incommensurability we are reflecting upon. However, we can argue that it is possible to understand the way we use social networks precisely through solipsism, narcissism, and incommensurability—we may understand the way we use social networks precisely through anarchism. Let us consider an example that is analogous to several other examples. When we post a video on YouTube, what we want is, outwardly, to actually share: in particular, to actually share a personal, and even intimate, experience with other people. Yet, what happens is the following:

1. We receive likes and dislikes from people we do not know, sometimes because we know them in an extremely superficial way and sometimes because we do not know them at all.
2. In any case, the likes we receive can affect us. We feel better, that is, happier and stronger, as soon as we see them and read analogous comments.
3. In any case, the dislikes we receive can affect us. We feel worse, that is, sadder and weaker, as soon as we see them and read analogous comments.
4. Thus, in order to optimize, as it were, our emotional state, we frequently use the following psychological strategy. On the one hand, we tell ourselves that, if people we do not know like us, then they are likely to be telling the truth, because they do not have any particular interest in saying that they like us, that is, people they do not know. On the other hand, we tell ourselves that, if people we do not know dislike us, then they are likely to be lying, because they have a particular interest in saying that they dislike us, that is, people they do not know: they are haters, that is, people who cure their own frustration by hating on other people through social networks.

If the phenomenon described is at least plausible, then we may say that what we do when we use social networks is not a matter of actually sharing personal, and even intimate, experiences with other people—inwardly, what we do seems to be a matter of preventing ourselves from actually sharing personal, and even intimate, experiences with other people—inwardly, what we do seems to be a matter of preventing ourselves from actually sharing at all.[3]

More precisely, we seem to use social networks, in particular, and technologies, in general, to try to establish a kind of autonomous sphere for ourselves. It is no coincidence that the etymological meaning of the word "autonomous" is analogous to the etymological meaning of "rulerless," since the former is the condition of possibility for the latter, which is its radicalization: a human being is "autonomous" if the "law" (νόμος) they follow is "self"-given (αὐτό-). Thus, if we are "autonomous," then we are not anarchistic, since we have a "law" (a rule), even though it is "self"-given. Yet, our use of social networks seems to show that we are moving from autonomy to anarchism. Indeed, there is another meaningful detail worth reflecting upon: most of the phases described and, in particular, most of the likes and the dislikes we receive are characterized by anonymity. And anonymity is what strengthens the psychological strategy described. On the one hand, we tell ourselves that, if anonymous people we do not know like us, then they are likely to be telling the truth, because they do not have any particular interest in saying that they like us, that is, people they do not know who are anonymous to them. On the other hand, we tell ourselves that, if anonymous people we do not know dislike us, then they are likely to be lying, because they have a particular interest in saying that they dislike us, that is, people they do not know who are anonymous to them: they are haters, that is, people who cure their own frustration by hating other people through social networks and, in particular, through the anonymity granted to them by social networks. Thus, we have people who are not only autonomous but also "autocratic," meaning, from an etymological perspective, that what is given to themselves by "themselves" (αὐτό-) is something meaningfully different from a law: it is a "power" (κράτος). More precisely, "autocracy" means the "absolute power of a monarch who depends solely on themselves, that is, who is not bound by any law."[4] Thus, the logical step from autonomy to anarchism via autocracy becomes clear: moving from autonomy to autocracy means moving from being bound by a law, even though it is self-given, to not being bound by a law at all—autocracy means that the law is replaced by the power to follow no rule at all: the stability of rule (which is stable both if it is heteronomous and if it is autonomous) is replaced by the instability of contingency, which means that we do anything we contingently want to do.

Let us go back to our use of social networks and, in particular, to our example. Paradoxically enough, when we outwardly share, we inwardly isolate ourselves through a frequent anonymity that neutralizes possible rules given not only by other people but also by ourselves, both when we read comments about us (in this case, we can use the psychological strategy that is contingently more reassuring for our emotional state) and when we write comments about other people (in this case, we can use the anonymity through which we write anything we contingently want to write to reassure ourselves when we feel frustrated). Thus, we isolate ourselves in a reassuring solipsism in which we can do any contingent and accidental thing we want to do.

Finally, the logical step from autocracy to anarchism becomes clearer: after the replacement of the law with the power to follow no rule at all, we are, again, "rulerless." And the only thing we follow is our idiosyncratic emotional state, which becomes weaker and weaker, since it increasingly stops to be trained to measure up to rules—anarchism is the laziest form of epistemology (and of ethics) precisely because it removes what may be thought of as humans' best trainer: the (ideal) rule to measure up to.

A SECOND CASE IN POINT: THE WEB, MONEY, AND POLITICS

Ethereum, a blockchain, is officially defined as "a global, open-source platform for decentralized applications,"[5] and, in particular, as not "controlled by any government or company—it is decentralized, [. . .] meaning that no single entity or person controls [it]. The Ethereum community is the largest and most active blockchain community in the world. It includes core protocol developers, cryptoeconomic researchers, cypherpunks, mining organizations, Ethereum holders, app developers, ordinary users, anarchists, fortune 500 companies, and, as of now, you. There is no company or centralized organization that controls Ethereum."[6] "[A]narchists" are more than officially included. Ethereum may be described:

> as a single, massively distributed computing engine sprawled across the global network, in which all processing is paid for in increments of Ether. In this model, computing power aggregated from that of individual machines becomes immanent to the network itself. This architecture allows Ethereum to serve as a platform for networked applications that run on a decentralised, peer-to-peer basis, rather than the more conventional client-server model. [...] it does confer at least one significant advantage for applications that skirt the law, like file-sharing: when functionality is smeared out across the entire network, no one jurisdiction can suppress it, however much they might want to. (Greenfield 2018: 149)

Interestingly enough, Ethereum may be described by adding to the meaningful word "anarchists" the following equally meaningful words:

1. "Decentralized" and "peer-to-peer basis," which mean that there is no ontological hierarchy distinguishing a superior dimension (for instance, the expert, as well as the controller) from an inferior dimension (for instance, the nonexpert, as well as the controlled).
2. Not "controlled by any government or company," "no single entity or person controls [it]," and "no company or centralized organization that controls Ethereum," which mean that there is a form of autocracy.
3. "Skirt the law" and "no one jurisdiction can suppress it," which mean that there is something more than a form of autocracy: indeed, there is a form of anarchism.
4. "Immanent to the network itself," which means that there is immanence—there is precisely what I am trying to argue, that is, a totally immanent technology characterized by the typically divine prerogatives of omnipresence, omniscience, omnipotence, and inscrutability (at least

> potentially) and, at the same time, by the total immanence that defines it as the ultimate result of the most radical form of anarchism we have ever experienced.

Moreover, Ethereum launched another technology, which was even more radical (and which finally failed): Distributed Autonomous Organization (DAO), which aimed at radicalizing the features of decentralized organizations, distributed applications, and autonomous agents. DAOs may be described as "nonhierarchical decision-making frameworks unsanctioned by any apparatus of state. In this oddball investment vehicle, some of the most dedicated community activists I've ever met glimpse the makings of an operating system for a planetary anarchism" (Greenfield 2018: 168). But, also in this case, the balance between forms of extremism, that is, anarchism and totalitarianism, is critical, as we have seen. In any case, a DAO has an originating owner, which "can add (or remove) voting members to the organisation" (Greenfield 2018: 168), which means that the originating owner "sets its binding decision rules, and retains the capacity to change that rule whenever it desires" (Greenfield 2018: 168) to "change the rules of participation at any time" (Greenfield 2018: 174). Moreover, anonymity via pseudonymity makes the balance between anarchism and totalitarianism even more critical: "DAO acts to diffuse and obscure accountability for the things it does. So long as pseudonymous participation is allowed by the bylaws of a DAO, we have no real way of assessing the beneficial ownership of that organization. We can inspect the blockchain at any time we please, identify the tokenised signatures of all its voting members. But we have no idea if that token denotes an individual, a group entity, or a chunk of code recursively delegated by some other DAO or autonomous process. With ownership screened behind a cryptographic baffle, the relationship between individuals, groups and entities in the world of DAOs is murky at best" (Greenfield 2018: 176). However, anarchism seems to win out over totalitarianism: the ultimate result is that "the relationship between individuals, groups and entities" dissolves into solipsism.

Anonymity via pseudonymity has another meaningful consequence: again, the ontological distinction between the human and the technological becomes indiscernible. The ontological distinction between, on the one hand, "individuals, groups" (i.e., humans and human institutions) and, on the other hand, "entities" (i.e., technologies) "is murky at best." Indeed, "we have no idea if that token denotes an individual, a group entity, or a chunk of code recursively delegated by some other DAO or autonomous process." Again, the indiscernibility between the human and the technological leads to anarchism: the ontological prerogatives of the technological and, in particular, its total immanence characterize the human not only by making it into a kind of autocratic monad that behaves being pushed by contingent need and will but also by making it into a kind of anarchic monad that is allowed to behave in any way it wants, including outlawed behaviors, by confusing its identity with technological identities. And the ultimate result is that the human being is unlikely to be found guilty, not only from a legal perspective but also, and especially, from an ethical perspective—again, the ultimate result is that the human being is unlikely to be defined as individually responsible: the more indiscernible the human and the technological become, the more the former obtains the latter's prerogative of being freed from individual responsibility, sometimes only ethically and sometimes both ethically and legally.

Interestingly enough, the quote from Adam Greenfield may be applied to another case in point. Speaking of an originating owner that "sets its binding decision rules, and retains the capacity to change that rule whenever it desires," including the capacity to "add (or remove) voting members to the organization," may bring to mind what is happening in Italy with the Five Star Movement, which uses technology, that is, a digital platform, to try to move from representative democracy to direct democracy.

Let us start by clarifying the meanings of the two kinds of democracy. We may summarize by saying that direct democracy is distinguished from representative democracy by the removal of the role of the expert as the mediator between people's incompetence and politicians' competence. In the case of representative democracy, when we outwardly delegate individual responsibility, we inwardly exercise the following individual responsibility: after having recognized our incompetence in a certain matter, we exercise the individual responsibility of choosing sometimes the party and sometimes the politician whose competence represents us. In the case of direct democracy, when we outwardly do not delegate individual responsibility, we inwardly do the following: after not having recognized our incompetence in a certain matter, we exercise the individual responsibility of choosing what do to through our vote, but our vote is likely to be more irresponsible than responsible since it is likely to be totally incompetent. For instance, if the issue requires a professional economic competence, and we choose what to do through our vote without having any professional economic competence, then we are irresponsible rather than responsible, because, again, we entrust ourselves, as well as our society, to extreme contingency—we entrust ourselves, as well as our society, to what is most extremely contingent, that is, our totally random idiosyncratic situations (economic incompetence included).

However, representative democracy is experiencing a crisis that, not coincidentally, is becoming characteristic of our (anarchic) technological era.[7] The Italian case of the Five Star Movement clearly exemplifies the crisis.[8] The reason why Greenfield's words may be applicable is that they highlight an ambiguity, that is, a critical balance between forms of extremism, which seems to characterize both the case of DAO and the case of the Five Star Movement.

Let us start by summarizing the most defining features of the Five Star Movement (see at least Biondo and Canestrari 2019; Corbetta and Gualmini 2013; Iacoboni 2018; Santoro 2012; Tronconi 2015):[9]

1. Its website states that "the Five Star Movement is a free association of citizens. It is not a political party, and it is not supposed to become a political party in the future. No left or right ideologies, but ideas. It wants to achieve an efficient and effective exchange of opinions and democratic debate outside associative and party ties and without the mediation of governing or representative bodies, recognizing to all citizens the governing and steering role normally attributed to few."[10]
2. In particular, it defines itself as post-ideological, anti-European, against party politics, and for direct democracy.[11]
3. When it entered the Italian Parliament, it did not follow a predetermined political view. On the contrary, its members of parliament voted on legislation on a case-by-case basis. Thus, they rejected stable political alliances with others (at least at the beginning).
4. Its rhetoric is quite violent. For instance, in 2007, it held the V-Day (Vaffanculo Day in Italian, Fuck You Day in English) against party politics and, in 2012, it organized an event called the Tsunami Tour.
5. When it entered the Italian Parliament, its political consultations were streamed online.
6. When it entered the European Parliament, its founder and guarantor (Beppe Grillo) gave its activists three predetermined options to choose from: first, joining the parliamentary group Europe for Freedom and Direct Democracy (EFDD), which won; second, joining the parliamentary group European Conservatives and Reformists (ECR); third, joining no parliamentary group.
7. It is against the public funding of political parties. On the contrary, it is funded both by a part of the salary of its members of parliament and by private donations.
8. It is against trade unions and for co-management by owners and workers.
9. It is for imperative mandate, which rejects one of the cornerstones of representative democracy by forcing the members of parliament to vote according to the program of the Five Star Movement (on the contrary, according to one of the cornerstones of representative democracy, as well as according to the Italian Constitution, members of parliament should represent, primarily, the general interests of the state and, secondarily, the particular interests of the party. Thus, there should not be any imperative mandate at all).

Before analyzing the relationship between the most defining features of the Five Star Movement and anarchism, let us add other meaningful details. In recent years, the Five Star Movement has experienced a crisis both in terms of votes, which have severely decreased from 2018 to 2020, and in terms of the relationship between its leadership and its members of parliament. In particular, they have frequently left the movement or been expelled by the leadership, together with other activists, from the movement. Interestingly enough, the members of parliament who joined other parties made extremely wide-ranging choices, from far-left parties to far-right parties. And, in most cases, they said that the reason why they left the movement, whether because of their free choice or because of the guarantor's unilateral decision, was the authoritarianism of the leadership.

Another meaningful detail is the relationship between representative democracy (which should be removed) and direct democracy (which should be established). First, as we have seen, there is no "mediation of governing or representative bodies, recognizing to all citizens the governing and steering role normally attributed to few." Second, the activists we may call party officials, occupying an intermediate position between the leadership and the other activists, are called, on the contrary, facilitators. Third, several activists, from the facilitators to the members of parliament, proved to be incompetent. Indeed, "recognizing to all citizens the governing and steering role normally attributed to few" means risking to elect incompetent citizens. Fourth, and most interestingly, what the Five Star Movement uses to try to move from representative democracy to direct democracy is technology and, in particular, digital technology (it is no coincidence that the word "e-democracy" is used to define, among other experiments, what the Five Star Movement is trying to do). A digital platform called Rousseau[12] (the philosopher considered by the founders an inspirer of direct democracy) allows the activists both to vote in any referendum launched by the leadership and to contribute to the drafting of laws. In particular, the activists vote in referenda as follows:

1. The activists must have been members of the movement for at least six months before voting in a referendum (membership requires an identity document), in order to prevent people extraneous to the movement from joining it at the last minute to sabotage the result of the referendum.
2. There are two different databases: the first contains the data on the voters (who are allowed to vote only once) and the second contains the data on the votes. As such, it is not possible to correlate the identities of the voters with the contents of the votes.
3. In any case, a notary certifies the result of the referendum.
4. The digital platform is funded both by a part of the salary of the members of parliament of the movement and by private donations.

Several details can lead us to argue that there is a meaningful correlation between the most defining features of the Five Star Movement and anarchism. Let us focus on the most meaningful details from a philosophical perspective. First, the role of the expert as the mediator between people's incompetence and politicians' competence is removed. Second, the role of the rule as the superior (universal, ideal) ontology that guides the inferior (particular, real) ontology of human affairs experiences a severe crisis.

As for the first issue (i.e., the removal of the role of the expert), we can highlight the following: first, according to what is written on the movement's website, the word "citizens" (repeated twice) replaces other words, such as "politicians," "representatives," and even "activists" (in particular, the word "citizens" replaces "the mediation of governing or representative bodies"); second, according to what is written on the movement's website, the word "all" replaces the word "few"; third, the co-management of owners and workers replaces trade unions; fourth, facilitators replace party officials; fifth, activists vote in complex referenda and contribute to drafting complex laws, even if they lack any professional competence; sixth, it is no coincidence that their political leader from 2017 to 2020,[13] with no degree or consolidated professional expertise, was the Minister of Economic Development, as well as the Minister of Labour and Social Policy, in 2018 (at the age of thirty-two), and the Minister of Foreign Affairs and International Cooperation in 2019 (at the age of thirty-three).

As for the second issue (i.e., the crisis of the role of the rule), we can highlight the following: first, according to what is written on the movement's website, the words "free associations" replace the words "political party"; second, according to what is written on the movement's website, the words "ideas" and "opinions" replace the word "ideologies" (interestingly enough, the word "opinions," "opinioni" in Italian, corresponds to the Greek word δόξα, meaning, from an epistemological perspective, the "subjective knowledge" that, being particular, contingent, and uncertain, opposes the Greek word ἐπιστήμη, meaning, from an epistemological perspective, the "objective knowledge" that is universal, absolute, and certain); third, according to what is written on the movement's website, being against "ideologies" means being indefinable, since the political categories, that is, "left or right," are replaced by the vague words "free association of citizens"; fourth, in parliament, a predetermined political view is replaced by a case-by-case basis, which means that the universality of the rule, which works as an asymptotic ideal model for the members of parliament, is replaced by the particularity of the case, which works as a contingent and random freedom for the members of parliament; fifth, a strong autocracy, which we can define as anarchism, shows up in terms of the party's rejection of both stable political alliances with others (at least at the beginning) and public funding, in order to get to a kind of solipsism founded on cutting all ties with both others and the state; sixth, streaming political consultations online means rejecting hierarchies, as well as the idea that there are complex issues that require a kind of exclusivity, from a matter of competence to a matter of confidentiality for security reasons.

Finally, there is the major issue of direct democracy replacing representative democracy. On the one hand, direct democracy may be considered close to anarchism in that, by removing the role of the expert as the mediator between people's incompetence and politicians' competence, incompetent people can be "rulerless" and, in particular, can be pushed by contingent need and will, as well as by totally random idiosyncratic situations. On the other hand, direct democracy may be considered close to totalitarianism in that, as we have seen, any form of extremism continuously risks to fall into its opposite form of extremism. As such, existing "without the mediation of governing or representative bodies" means continuously risking to fall into the kind of authoritarianism we have seen, for instance, in terms of giving the activists three predetermined options to choose from (when the movement entered the European Parliament), giving the members of parliament an imperative mandate and expelling the members of the parliaments and other activists because of the guarantor's unilateral decision.

It is no coincidence that, if we go back to a possible contemporary Prometheus, and in particular to Eggers' novel, then we can find a fairly illuminating analogy between the kind of direct democracy founded on, for instance, streaming online and Eggers' "100 percent democracy" (Eggers 2014: 390), which starts with a politician who agrees to wear a camera at all times, including when she is managing the most sensitive issues, and which ends up in a program with the following aims: "eliminate lobbyists. Eliminate polls. It might even eliminate congress" (Eggers 2014: 395), together with what we have seen, that is, "everyone will be tracked, cradle to grave, with no possibility of escape." Again, what we may define as a technological panopticon can equally underpin both forms of anarchism (starting from allowing everyone to see everyone else) and forms of totalitarianism (starting from allowing one person to see everyone else). And, again, it may be a short step from the former to the latter.

There is at least one other interesting correlation between the most defining features of the Five Star Movement and one of the most typical consequences of epistemological anarchism: the risk to fall into the phenomenon of fake news. On May 2, 2017, a *New York Times* editorial titled *Populism, Politics, and Measles*[14] stated that "In Italy, the populist Five Star Movement (M5S) led by the comedian Beppe Grillo has campaigned actively on an anti-vaccination platform, likewise repeating the false ties between vaccinations and autism. To these and other skeptics, the measles outbreak in Italy should sound a piercing alarm. As of April 26, the Italian Ministry of Health had reported 1,739 cases of the disease, compared with 840 in all of 2016 and only 250 in 2015. Of those stricken, 88 percent had not been vaccinated. The danger was not only to them: 159 of the cases were health care workers infected by patients. Yet studies show that 97 percent of people who receive the recommended two doses of MMR (measles, mumps and rubella) vaccine are fully protected. Most people today would not remember a time when measles—or mumps, or polio—were commonplace. M5S may not be responsible for the entire outbreak, since vaccine skepticism predates the party's rise. Yet, the percentage of two-year-olds given vaccinations has steadily fallen in recent years, from 88 percent in 2013 to 86 percent in 2014 and 85.3 percent in 2015. The World Health Organization regards 95 percent as the level to achieve 'herd immunity,' at which point the disease poses no threat to the entire community." The editorial started with the following words: "One of the tragedies of these post-truth times is that the lies, conspiracy theories, and illusions spread by social media and populist politicians can be downright dangerous." We may add to these words, which sound shocking as we face the coronavirus emergency, that, if "populist politicians" and "social media" are strengthened by sophisticated technologies, as in the case of the Five Star Movement, then the risk to fall into "post-truth" exponentially increases. And "post-truth" is, again, correlated with epistemological anarchism. "Post-truth," defined, not coincidentally, as the word of the year by the Oxford English Dictionary in 2016, refers to "circumstances in which objective facts are less influential in shaping political debate or public opinion than appeals to emotion and personal belief."[15] And "emotion and personal belief" correlate, on the one hand, with the idiosyncratic emotional state and, on the other hand, with the solipsistic, contingent, and accidental truth that, as we have seen, characterize epistemological anarchism, ending up replacing truth with "post-truth": "'Truth', written 'in capital letters', is an orphan in this world, without power and influence" (Feyerabend) and "There is no final Truth" (Tzara)—we can argue that "post-truth" refers to anarchism especially, first, in terms of removing the role of the expert as the mediator between ignorance and truth, as well as between competence and incompetence, and, second, in terms of removing truth itself, which ends up being something literally "post" what it has been for millennia for Western humans.

In this second paragraph of this third chapter, we have analyzed, first, a case of digital technology applied to money and, second, a case of digital technology applied to politics. There are several analogous examples that cannot be listed in a book that is not focused on collecting them but on their philosophical meaning. We can argue that they have two essential philosophical meanings, both strictly correlating with anarchism.

Their first essential philosophical meaning is epistemological, strictly correlating with epistemological anarchism. We are progressively stopping to exercise our ability to make decisions by ourselves by referring to a challenging epistemological ideal. We stop to challenge an epistemological ideal, as well as an epistemological hierarchy, in the case of money, especially using a "peer-to-peer basis" and a "power" that is "immanent to the network itself" and, in the case of politics, especially by replacing "ideologies" with "opinions," as well as with fake news, by replacing a predetermined political view with a case-by-case basis and also with no professional competence, by voting in complex referenda, and by contributing to drafting complex laws.

Their second essential philosophical meaning is ethical, strictly correlating with ethical anarchism. We are progressively stopping to exercise our ability to bear the burden of responsibility for our decisions by referring to a challenging ethical ideal.[16] We stop to challenge an ethical ideal, as well as an ethical hierarchy, in the case of money, especially by not being "controlled by any government or company," by "skirt[ing] the law", and by replacing individual responsibility with anonymity via pseudonymity and, in the case of politics, especially by rejecting, first, stable political alliances (at least at the beginning), second, hierarchies via streaming online, and, third, public funding, that is, by getting to a kind of solipsism founded on cutting any relationship with both others and the state.

Since the end of the eighteenth century, we, as Western humans, have been learning that we are essentially free to choose for ourselves in an autonomous way. Indeed, Kant's philosophy in particular (see especially Kant 1788), as well as the Enlightenment culture in general, together with the French Revolution, underpinned the idea that every human being is essentially free to choose for themselves autonomously. Thus, we think of ourselves as essentially free to choose, for instance, to be whoever we want to be—including to be the exact opposite of the kind of humans we have been for about two and a half millennia, when we, as Western humans, started to think of ourselves as the kind of humans who are essentially, first, epistemological decision-makers (by referring to a challenging epistemological ideal) and, second, ethical decision-makers (by referring to a challenging ethical ideal), and even courageous bearers of the burden of responsibility for our decisions.

Thus, we are free to choose to be the exact opposite of the kind of humans we have been throughout Western history. But we should be conscious of what we are doing: it may be the most radical revolution we have ever experienced as Western humans, who choose to be anarchistic humans who externalize (as we shall see in the fourth chapter) epistemological and ethical decisions, as well as the burden of responsibility for them, by creating a totally immanent technological divine that makes epistemological and ethical decisions for us—freeing us from bearing the burden of responsibility for any decision, and even for any responsibility whatsoever.

This may be the most radical revolution we have ever experienced as Western humans, since we have been thinking of ourselves for millennia as follows: "Who more miserable than I, what mortal could you find more god-abhorred? Wretch whom no sojourner, no citizen may harbour or address, whom all are bound to harry from their homes. And this same curse was laid on me, and laid by none but me. Yea with these hands all gory I pollute the bed of him I slew. Say, am I vile? Am I not utterly unclean, a wretch doomed to be banished, and in banishment forgo the sight of all my dearest ones, and never tread again my native earth; or else to wed my mother and slay my sire, Polybus, who begat me and upreared? If one should say, this is the handiwork of some inhuman power, who could blame his judgment? But, ye pure and awful gods, forbid, forbid that I should see that day! May I be blotted out from living men ere such a plague spot on me its brand!" (Soph. *Oed. King* 815–33). As Sophocles teaches us through his outstanding figure of Oedipus, we, as Western humans, have been thinking of ourselves as able to bear the burden of responsibility and, therefore, able to risk to tragically feel "miserable," "god-abhorred," "vile," "utterly unclean," "doomed to be banished," and "blotted out from living men," even when it has been a matter of predestination. And, when, after several centuries, it has not been a matter of predestination, the burden of responsibility we have been bearing has become heavier, as William Shakespeare's equally outstanding figure of Hamlet teaches us: "For who would bear the whips and scorns of time, th' oppressor's wrong, the proud man's contumely, the pangs of despis'd love, the law's delay, the insolence of office, and the spurns that patient merit of th' unworthy takes, when he himself might his quietus make with a bare bodkin? Who would fardels bear, to grunt and sweat under a weary life, but that the dread of something after death, the undiscovered country, from whose bourn no traveller returns, puzzles the will, and makes us rather bear those ills we have than fly to others that we know not of? Thus conscience does make cowards of us all, and thus the native hue of resolution is sicklied o'er with the pale cast of thought, and enterprises of great pitch and moment with this regard their currents turn awry, and lose the name of action" (Shakespeare. *Hamlet.* 1776–94). However, we have kept training our ability to bear the burden of responsibility for our decisions, even when, after centuries, it has become heaviest, being the burden of an individual identity that becomes indiscernible, as Luigi Pirandello, the Nobel Prize winner for literature in 1934, shows us through "the discovery of the hundred thousand Moscardas that I was, not only for the others but also for myself, all with this one name of Moscarda, ugly to the point of cruelty, all inside this poor body of mine that was also one" (Pirandello 1992: 14). Thus, "all the reality surrounding you has, for the others, no more substance than that smoke" (Pirandello 1992: 25): even though "We used, you and I, the same language, the same words" (Pirandello 1992: 31), "you fill them with your meaning, as you say them to me; and I, as I hear them, inevitably fill them with my meaning. We thought we understood each other; we didn't understand each other at all" (Pirandello 1992: 31–2). Finally, we can find what marks, from an epistemological perspective, the crisis of the classical relationship between ideality's universality and reality's particularity in the Western culture (and it is no coincidence that Pirandello's novel was published in 1926): "If only there were outside of us, for both, a my reality and a your reality, both sovereign, I mean in themselves, and at the same time, unchangeable. There isn't. There is in me and for me a my reality: the one I give myself; a your reality in you and for you, the one you give yourself; they will never be the same. So, my friend, we must console ourselves with this thought: mine is no more true than yours, and they last only a moment, both yours and mine" (Pirandello 1992: 32). The words that mark Pirandello's reflection are somehow analogous to the words that mark the special form of epistemological anarchism we are reflecting upon, in that the former somehow anticipates the latter. More precisely, we can find the anti-foundationalism (there is no "unchangeable" reality shared by all humans) that typically results in, first, epistemological solipsism (there are "a my reality" and "a your reality" that "will never be the same"), second, epistemological relativism ("mine is no more true than yours"), and, third, epistemological contingency ("they last only a moment")—which means anticipating epistemological anarchism. From a logical perspective, speaking of the dissolution of an epistemological ground that is thought to be "unchangeable" enough to be sharable between more than one individual, both synchronically and diachronically, means speaking of the dissolution of the very condition of possibility for having something that rules at all, being the "unchangeable" ideal rule of changeable real accidents, since everything "last[s] only a moment."

But what we are doing in our technological era is even more radical: we are not only depriving our decisions of rules "unchangeable" enough to be sharable but also depriving ourselves of making our own decisions at all (as we shall discuss in the fourth chapter).

NOTES

1. See for instance *Coronavirus, l'anestesista di Codogno che ha intuito la diagnosi di Mattia: "Ho pensato all'impossibile,"* "La Repubblica," March 6, 2020 (https://www.repubblica.it/cronaca/2020/03/06/news/l_anestesista_di_codogno_per_

mattia_era_tutto_inutile_cosi_ho_avuto_la_folle_idea_di_pensare_al_coronavirus_-250380291/).

2. See at least Finn (2017) for an analysis from a philosophical perspective.

3. This issue is addressed by several disciplines, from the humanities to the social sciences, from medicine to psychology. References are too numerous to be listed.

4. See https://www.etimo.it/?term=autocrazia (my translation).

5. Quote from https://ethereum.org/.

6. Quote from https://ethereum.org/what-is-ethereum/.

7. References are several and from several disciplines. See at least Alonso, Keane, Merkel and Fotou (2011); Crouch (2004); Dalton (2012); Della Porta (2013); Keane (2009); Rosanvallon (2008).

8. Together with other European movements, at least partially: for instance, Pirate Parties and Movimiento 15-M.

9. Beppe Grillo (a comedian and activist) and Gianroberto Casaleggio (a digital entrepreneur) officially founded it in 2009. The latter died in 2016 (with his son Davide Casaleggio taking his place), and the former exercises the role of guarantor.

10. Quote from https://www.movimento5stelle.it/ (my translation).

11. See also https://www.ilblogdellestelle.it/.

12. See https://vote.rousseau.movimento5stelle.it/.

13. Luigi Di Maio.

14. *Populism, Politics, and Measles*, "The New York Times," May 2, 2017 (https://www.nytimes.com/2017/05/02/opinion/vaccination-populism-politics-and-measles.html.

15. Quote from https://www.oed.com/view/Entry/58609044?redirectedFrom=post-truth#eid.

16. Most interestingly, according to the Moral Machine experiment conducted by the Massachusetts Institute of Technology, the Western cluster shows the highest preference for inaction (the other clusters are the Eastern one and the Southern one). See E. Awad *et al.*, *The Moral Machine Experiment*, "Nature," n. 563 (2018), pp. 59–75. I shall discuss the issue of epistemological and ethical discharging in the following chapters.

Chapter 4

Radical Externalization

THE MOST RADICAL EPISTEMOLOGICAL EXTERNALIZATION AND META-EXTERNALIZATION

The third chapter ended with a quote from Hamlet's monologue. Let us start the fourth chapter by quoting another Hamlet's monologue, which was written by a philosopher about three centuries after Shakespeare's masterpiece. Surprisingly enough, the philosopher is the rigorous logical empiricist Hans Reichenbach. His Hamlet asks himself the following:

> To be or not to be—that is not a question but a tautology. I am not interested in empty statements. I want to know the truth of a synthetic statement: I want to know whether I shall be. Which means whether I shall have the courage to avenge my father? [. . .] I have good evidence. The ghost was very conclusive in his arguments. But he is only a ghost. Does he exist? [. . .] But that's it: nothing but indirect evidence. Am I allowed to believe what is only probable? Here is the point where I lack the courage. [. . .] I am afraid of doing something on the basis of a mere probability. [. . .] what if I should start thinking after the deed and find out I should not have done it? [. . .] the logician [. . .] tells me that if something is probable I am allowed to make a posit and act as though it were true. In doing so I shall be right in the greater number of cases. But shall I be right in *this* case? (Reichenbach 1959: 250–1)

Reichenbach's Hamlet ends up answering: "There is no certainty. The probability will be increased and my posit will have a higher rating. I can count on a greater percentage of correct results. That is all I can reach. I can't get away from making a posit. I want certainty, but all the logician has for me is the advice to make posits. There I am, the eternal Hamlet. His advice confirms my doubt rather than giving me the courage I need for my action" (Reichenbach 1959: 251).

I think that these lines are exceedingly meaningful, as well as touching, precisely because their author is not Shakespeare himself, but one of the most rigorously rational philosophers of Western philosophy. We may read them as touching explanations of the reason why Reichenbach's Hamlet and, symbolically, every human being (at least in the Western culture) seem to be destined to be an "eternal Hamlet"—and the reason is that no human life seems to be free of the need to act in critical situations, that is, when our action cannot be founded on any certainty, even though it has the potential for dramatic, and even tragic, consequences on both ourselves and others.

More precisely, we can analyze Reichenbach's lines as follows:

1. We must act, and our action has a dramatic consequence.
2. Thus, in order to act, we need an ethical condition: we need the courage to act (in Reichenbach's Hamlet's case, "the courage to avenge my father," "the courage I need for my action," but "I lack the courage," "I am afraid").
3. Thus, in order to have the courage to act, we need an epistemological condition: we need certainty (in Reichenbach's Hamlet's case, "I want to know the truth" and "I want certainty" on "*this* case").
4. But we have no certainty at all (in Reichenbach's Hamlet's case, "there is no certainty" and we have "nothing but indirect evidence," something that "is only probable" and "a mere probability").
5. Thus, in order to have certainty, we ask the logician, but they tell us that the only thing we can do is "to make posits." From a logical perspective, "if something is probable I am allowed to make a posit and act as though it were true," since "I shall be right in the greater number of cases."
6. Thus, we have no certainty at all. We have our "posit," but it cannot answer our question: will we "be right in *this* case?"
7. Thus, we are destined to be an "eternal Hamlet," whose doubt remains strong (the logician's "advice confirms my doubt") and, therefore, whose courage to act remains weak ("rather than giving me the courage I need for my action").

Yet, we must act. But what can we do?

I can answer this question by anticipating that I believe that the special form of technological anarchism we are experiencing in our era seems to provide us with a possible way out.

Reichenbach's words are meaningful also because of his choice of the word "posit," whose Latin etymology means "to take" (*sumere*) something "upon" (*ad*) ourselves. Thus, from an epistemological perspective, making a "posit" means that what we do when we have no certainty to rely on is to rely only on what is "taken upon" ourselves. And what is "taken upon" ourselves is by definition characterized as follows:

1. It is as particular, solipsistic, contingent, and accidental as we are.
2. It weighs, being, even etymologically, "taken upon" ourselves.

Thus, we find ourselves in a paradoxical situation, which is doubly disadvantageous for us: on the one hand, we have no certainty (first point) and, on the other hand, we bear a burden (second point). Interestingly enough, we meet again what we have seen in the third chapter, that is, the idea of bearing a burden, whose meaning becomes clearer: we bear the burden of responsibility for our actions founded on our totally uncertain decisions, which can have dramatic, and even tragic, consequences on both ourselves and others.

I believe that our era may be read as our greatest struggle to try to shift this burden from ourselves to technology through the most radical form of anarchism we have ever experienced precisely because it can neutralize Reichenbach's Hamlet's unbearable question: "But shall I be right in *this* case?"

Thus, we are increasingly externalizing everything, from every epistemological decision to every ethical decision—more precisely, we are increasingly shifting our decisions from our minds, as well as from our bodies, to the algorithmic machines we create.

Let us start with a metaphor. Let us imagine that we want to get to the top of a mountain and that we have two different ways of doing so at our disposal: the first is climbing the mountain and the second is going by helicopter. It goes without saying that both cases require technologies, even though they are different—as we have learned from the very genesis of the Western culture, the human being can be defined as the technological being, who cannot do without technology at all. But we can choose different technologies, of course. In our example, the first case requires a great human effort, since we need to use our physical strength, as well as our mental strength, in order to control every muscle, as well as every decision on what move to make, whereas the second case requires little effort (if we are the passengers and not the pilots). In both cases, we have both advantages and disadvantages:

1. When we climb the mountain, we have at least the advantages of becoming physically and mentally stronger and feeling the pleasure of doing something mostly with our own strength. And we have at least the disadvantages of getting to the top of the mountain later and being exhausted once we get there.

2. When we go by helicopter, we have at least the advantages of getting to the top of the mountain sooner and feeling fresh, that is, immediately ready to experience the pleasure of contemplating the beautiful view. And we have at least the disadvantages of not becoming physically and mentally stronger (since we are not training any of our abilities) and not feeling the pleasure of doing something mostly with our own strength (since we are exclusively using the power of the helicopter). Thus, what is the best choice? This is hard to answer.

But there is a meaningful detail worth reflecting upon. Let us ask ourselves what we delegate in both cases—more precisely, what we externalize by shifting something from ourselves to technology. We should answer that, in both cases, we do not externalize our aim: we want to get to the top of the mountain and, in both cases, we get to the top of the mountain. But we should also answer that, in the first case, we do not radically externalize the way we do it (using our own strength), whereas, in the second case, we radically externalize the way we do it (using the power of the helicopter). More precisely, we should say, in the first case, that we do not radically externalize anything, that is, any content or abilities, whereas, in the second case, that we do not radically externalize any content, but we radically externalize abilities.

Speaking of content means speaking of our aim (i.e., getting to the top of the mountain): in both cases, our aim is satisfied. Speaking of abilities means speaking of our way of satisfying our aim (i.e., are we using an internal strength? Are we using an external power?). In the first case, our way of satisfying our aim is using our own strength, whereas, in the second case, our way of satisfying our aim is using the power of the helicopter. Thus, we can argue that there are two kinds of delegation—there are two kinds of externalization: externalizing content, which we may define as an externalization (1) and externalizing abilities, which we may define as a meta-externalization (2).

Let us develop the argument. Any technology may be thought of as a way (2) of satisfying an aim (1). We frequently risk to neglect the issue of the way (2), which can be essential. For instance, I use several technologies as ways (2) of satisfying my aim of remembering what I have to do (1), from handwriting to digital writing. When I choose to use these technologies, I choose not to use my memory (and I can honestly say that my memory is quite out of training: I can hardly remember what I have to do today if I do not read what I wrote). Thus, also in my case, I can distinguish one aim and two ways: the aim of remembering what I have to do (1), which does not change, and the ways of satisfying it (2), which change, being memory in my past and handwriting and digital writing in my present. In my past, I could perfectly satisfy my aim. In my present, I can perfectly satisfy my aim. But I know that something essential has changed over the years: my memory has weakened.

Thus, saying that any technology may be thought of as a way (2) of satisfying an aim (1) means saying that any technology may be thought of as something that keeps satisfying the aim (1) precisely by changing the way (2)—any technology may be thought of as something that keeps satisfying the aim (1) precisely by replacing a (mostly) human way with a (mostly) technological way. I know that any time I use handwriting and digital writing my memory weakens, but I decide that I am fine with it. On the contrary, if I go back to the example of getting to the top of the mountain, then I think the opposite. I know that any time I use the helicopter my strength weakens, and I decide that I am not fine with it at all. Why? Because I decide that something that is essential for my own identity as a human being is precisely my strength and, in particular, my mental strength (among other human abilities), but not my memory—I believe that this is the kind of decision that we, as human beings, should always make consciously: indeed, the reason why we should not neglect the issue of the way (2) is that deciding what ways to keep and what ways to meta-externalize means deciding precisely what kind of identities we want to have as human beings.

I may rephrase this argument as follows: meta-externalizing (abilities) is more crucial than externalizing (content) when it comes to deciding, in general, who a human being is and, in particular, who I am, who you are, etc.

Of course, you and I can decide differently. But it is important to try to share at least a reflection upon the decisions that severely impact us as a society. Thus, going back to the examples of getting to the top of the mountain and of remembering what I have to do, we should try to answer the following question: even though our decisions may be different (for instance, I decide to keep my strength and to meta-externalize my memory and you decide to keep your memory and to meta-externalize your strength), is there anything we can share? I think that we can share at least what may be the most crucial thing. If you and I decide to meta-externalize everything, that is, both our strength and our memory, as well as every other ability of ours, then we are likely to do the following:

1. Cause our abilities to atrophy and, therefore, become as powerless as the pre-technological ant-like humans described by Aeschylus, who "dwelt beneath the ground like swarming ants, in sunless caves."—most interestingly, meta-externalizing everything from the human to the technological is likely to cause the former to de-evolve into a pre-technological ontological status (as we shall see in the fifth chapter).
2. Deprive ourselves of the opportunity to experience our abilities, which, again, are crucial when it comes, first, to understanding, and, second, to deciding, who we are as human beings. For instance, when I use my physical and mental strength, I can understand what my limits are and

> what my powers are, that is, what it means to be myself. Moreover, I can decide what limits I want to decrease or not and what powers I want to increase or not, that is, what kind of identity I want to have as a human being—finally, what kind of life plan I want to have as a human being.

More precisely, I think that we can share the following idea: the most crucial reason why a radical meta-externalization of our abilities is likely to be a critical choice is not only a matter of neutralizing our abilities but also, and especially, a matter of neutralizing our understanding, our decision-making, and our planning of our identities as human beings precisely through our experiencing our abilities.

Thus, the issue of the way, that is, experiencing our abilities or not, can be essential precisely because it is nothing less than the issue of human identity—and we can summarize by arguing that meta-externalizing abilities is more crucial than externalizing content precisely because it is the former what can train us to evolve, which is the human ability *par excellence*: the ability to face especially what is most challenging, that is, the exceptions of the present and the surprises of the future.

However, we are massively meta-externalizing our abilities from ourselves to technology. In particular, we are massively meta-externalizing our ability to make decisions from ourselves to algorithmic machines (see at least Domingos 2015; Dyson 1997; Finn 2017; Garreau 2006; Harari 2018; Noble 2018; Pasquale 2015). As we have learned from the case of the "predictive algorithm" in the third chapter, the externalization of our most challenging puzzles may be extremely freeing. Thus, we are massively externalizing not only our decisions on health (from what tests we are going to take if we are patients to what tests we are going to prescribe if we are doctors) but also our decisions on relationships (for instance, via an algorithm that tells us who we are going to meet if we go out), on time spent alone (for instance, via an algorithm that tells us what film we are going to watch if we stay home), etc.—we are massively externalizing the very core of living human lives as they have been lived for millennia, from Sophocles to Shakespeare, from Pirandello to Reichenbach, as we have seen, that is, as a matter of being courageous bearers of the burden of responsibility for our decisions.

We have reflected upon the critical choice to radically meta-externalize our abilities in terms of neutralizing our ability *par excellence*—our ability to evolve, which is trained any time we experience our other abilities, especially the most challenging ones, such as the ability to make decisions. But we should also reflect upon this critical choice in the following terms: are we sure that the decisions made by algorithms are always better than our own?

THE MOST RADICAL HYPERTROPHIZATION OF *LOGOS*

I believe that one of the most promising ways of trying to answer this question may be reflecting not only upon our present but also upon our past and, in particular, upon our cultural history. Indeed, from a historical perspective, we can see a fairly precise trajectory, which, from our past to our present, seems to define an equally precise phenomenon, epistemological in particular and cultural in general, which we may call hypertrophization of *logos*—more precisely, a progressive hypertrophization of *logos* that, not coincidentally, seems to be experiencing its most radical form in our technological era.

In the following pages, I shall try to summarize, at least, the major steps of the history of this phenomenon, which constitutively characterizes Western epistemology in particular and the Western culture in general. By seeing this precise trajectory, I believe that we shall also be able to see what is happening in our technological era from a fairly illuminating perspective.

The first major step has to do, of course, with the genesis of Western philosophy. In particular, Plato argues in his *Theaetetus*:

> "let us not on that account lose sight of the question before us, which is: what is meant by the doctrine that the most perfect knowledge arises from the addition of rational explanation [*logos*] to true opinion?". "No, we must not." "Now what are we intended to understand by rational explanation? I think it means one of three things." "What are they?". "The first would be making one's own thought clear through speech by means of verbs and nouns, imaging the opinion in the stream that flows through the lips, as in a mirror or water." [. . .] "[The second would be] that each person if asked about anything must be able in reply to give his questioner an account of it in terms of its elements." [. . .] "[The third would be] the ability to tell some characteristic by which the object in question differs from all others. [. . .] As an example, if you like, take the sun: I think it is enough for you to be told that it is the brightest of the heavenly bodies that revolve about the earth. [. . .] Understand why I say this. It is because, as we were just saying, if you get hold of the distinguishing characteristic by which a given thing differs from the rest, you will, as some say, get hold of the definition or explanation of it; but so long as you cling to some common quality, your explanation will pertain to all those objects to which the common quality belongs. [. . .] Then he who possesses right opinion about anything and adds thereto a comprehension of the difference which distinguishes it from other things will have acquired knowledge of that thing of which he previously had only opinion. [. . .] Assume that I have right opinion about you; if I add the explanation or definition of you, then I

> have knowledge of you, otherwise I have merely opinion. [. . .] But explanation was, we agreed, the interpretation of your difference." (Plat. *Thaet.* 206 c–209 a)

Plato's direction is clear: the notion of *logos*, that is, "rational explanation," has to do with the notion of abstraction, that is, one of the most defining features distinguishing Western epistemology (see the second chapter). Indeed, Western epistemology is founded, first, on abstraction, by which we can identify what real objects share, and, second, on idealization, by which we can get to their ideal model. Finally, we use their ideal model as our guide in every cultural sphere, from science and technology to art, from law to religion. In particular, Plato speaks, first, of the ability to analyze, which underpins the ability to abstract (we may define analyzing as giving "an account of it in terms of its elements"), and, second, of the ability to abstract. Indeed, we may define the process of abstraction, which we have defined as the process of identifying what is shared by distinguishing it from what is not shared (see the second chapter), as getting "hold of the distinguishing characteristic by which a given thing differs from the rest": speaking of the distinction between what does not differ from what differs means speaking of the distinction between what is shared from what is not shared, which means, again, speaking of abstraction. Thus, the notion of *logos*, as well as the notions of abstraction and idealization, is nothing less than one of the most defining features distinguishing Western epistemology. And it also clarifies something essential: speaking of *logos* means speaking, again, of moving from the particularity of reality to the universality of ideality. In Plato's words, speaking of *logos* means speaking of getting "hold of the definition or explanation of it," and "the definition or explanation" is a matter of moving from what is particular to what is universal, that is, "the definition or explanation" that goes for everyone—and that works as everyone's privileged way to knowledge (in Plato's words, "if I add the explanation or definition of you, then I have knowledge of you, otherwise I have merely opinion").

Plato's stress on *logos* as our privileged way to knowledge is further strengthened by Aristotle, who writes in his *Nicomachean Ethics*:

> We are looking for the function peculiar to man; we must therefore set aside the vital activity of nutrition and growth. Next in the scale will come some form of sentient life; but this too appears to be shared by horses, oxen, and animals generally. There remains therefore what may be called the practical life of the rational part of man. (This part has two divisions, one rational as obedient to principle, the others possessing principle and exercising intelligence). [. . .] we declare that the function of man is a certain form of life, and define that form of life as the exercise of the soul's faculties and activities in association with rational principle. (Arist. *Nic. Eth.* I, 12, 1097 b–1098 a)

Thus, again through analysis and abstraction (i.e., the distinction between what is shared by every animal, meaning "nutrition and growth," as well as "sentient life," and what is not shared by every animal, meaning "the rational part"), Aristotle makes *logos* be the most defining feature distinguishing humans, from their reasoning to their speech (see especially his *The Art of Rhetoric*).

Logos becomes even stronger with the scientific revolution. In particular, Galileo's *Dialogue Concerning the Two Chief World Systems Ptolemaic and Copernican* is a major step of the history of what we may call the hypertrophization of *logos*. He not only works on the foundation of the modern scientific method by stressing the role of the most important result of *logos*, that is, the mathematical universal ideal model, which works together with empirical experience, but also compares human *logos* with divine intelligence: "our understanding, as well in the manner as in the number of things understood, is infinitely surpassed by the Divine; but I do not thereby abase it so much as to consider it absolutely null. No, when I consider what marvelous things and how many of them men have understood, inquired into, and contrived, I recognize and understand only too clearly that the human mind is a work of God's, and one of the most excellent" (Gal. *Dialog*. I). Most interestingly, we can recognize in these words the major issue of the historical evolution of the figure of Prometheus, that is, what we may call the theological challenge through science and technology: the human challenges the divine through science and technology, in that the former, first, has something of the divine ("the human mind is a work of God's and one of the most excellent") and, second, with "the human mind" has "understood, inquired into" (through science) "and contrived" (through technology). *Logos* is exceedingly important—*logos*, and in particular the ability of "the human mind" to mathematize, is not only the condition of possibility for science and technology but also the very core of human identity. And the theological challenge becomes even stronger: "taking man's understanding intensively, in so far as this term denotes understanding some proposition perfectly, I say that the human intellect does understand some of them perfectly, and thus in these it has as much absolute certainty as Nature itself has. Of such are the mathematical sciences alone; that is, geometry and arithmetic, in which the Divine intellect indeed knows infinitely more propositions, since it knows all. But with regard to those few which the human intellect does understand, I believe that its knowledge equals the Divine in objective certainty, for here it succeeds in understanding necessity, beyond which there can be no greater sureness" (Gal. *Dialog*. I). Galileo's argument is the following:

1. There is something human that can "understand [. . .] perfectly."

2. "Understand [. . .] perfectly" means that this kind of human understanding can equal divine understanding: "its knowledge equals the Divine in objective certainty," "it has as much absolute certainty as Nature itself has." In particular, "understand [. . .] perfectly" means that this kind of human understanding can move from contingency to "necessity," that is, from empirical experience to the mathematical universal ideal model, "beyond which there can be no greater sureness."
3. This kind of human understanding is given by "the human intellect," which is able to use "the mathematical sciences," that is, "geometry and arithmetic."

The ultimate result is the following: the most defining feature distinguishing the human being is "the human intellect," and the most defining feature distinguishing "the human intellect" is its ability to mathematize—what can be thought of as the most important result of the scientific revolution, at least from an epistemological perspective, is that *logos* becomes both more essential and more focused: more essential in that it is the very core of human identity and, in particular, its best part, and more focused in that it is not a matter of human understanding in general, but a matter of the ability of "the human intellect" to mathematize in particular.

It is no coincidence that Kant writes in the preface to the second edition of his *Critique of Pure Reason*:

> When Galileo rolled balls of a weight chosen by himself down an inclined plane [. . .] a light dawned on all those who study nature. They comprehended that reason has insight only into what it itself produces according to its own design; that it must take the lead with principles for its judgments according to constant laws and compel nature to answer its questions, rather than letting nature guide its movements by keeping reason, as it were, in leading-strings; for otherwise accidental observations, made according to no previously designed plan, can never connect up into a necessary law, which is yet what reason seeks and requires. Reason, in order to be taught by nature, must approach nature with its principles in one hand, according to which alone the agreement among appearances can count as laws, and, in the other hand, the experiments thought out in accordance with these principles—yet in order to be instructed by nature not like a pupil, who has recited to him whatever the teacher wants to say, but like an appointed judge who compels witnesses to answer the questions he puts to them. (Kant 1781: B 12–13)

The meaning of Kant's words is crucial: his Copernican revolution means that the human "reason" has in itself the "principles" required to get to "a necessary law." And its best tool to do it is, again, mathematics, which underpins science and technology: "Mathematics has, from the earliest times to which the history of human reason reaches, in that admirable people the Greeks, travelled the secure path of a science" (Kant 1781: B 10).

Before Kant's words on Galileo's scientific revolution, as well as before Kant's Copernican revolution, which is equally revolutionary from an epistemological perspective, there was at least one other philosopher who marked another major step of the history of the hypertrophization of *logos*: René Descartes, who was both a philosopher and a mathematician. His philosophy can be considered a radicalization of the idea that the very core of human identity is thought and, in particular, rational thought. Moreover, the very core of human identity as rational thought does not need a heteronomous foundation, since it is autonomously founded—and this is precisely the genesis of the Western focus on autonomy that leads to the Western focus on autocracy that, finally, leads to the Western focus on anarchism, as we have seen in the third chapter. Descartes writes in his *A Discourse on Method*:

> As I then desired to give my attention solely to the search after truth, I thought that [. . .] I ought to reject as absolutely false all opinions in regard to which I could suppose the least ground for doubt, in order to ascertain whether after that there remained aught in my belief that was wholly indubitable. Accordingly, seeing that our senses sometimes deceive us, I was willing to suppose that there existed nothing really such as they presented to us; and because some men err in reasoning, and fall into paralogisms, even on the simplest matters of geometry, I, convinced that I was as open to error as any other, rejected as false all the reasonings I had hitherto taken for demonstrations; and finally, when I considered that the very same thoughts (presentations) which we experience when awake may also be experienced when we are asleep, while there is at that time not one of them true, I supposed that all the objects (presentations) that had ever entered into my mind when awake, had in them no more truth than the illusions of my dreams. But immediately upon this I observed that, whilst I thus wished to think that all was false, it was absolutely necessary that I, who thus thought, should be somewhat; and as I observed that this truth, I think, therefore I am (*cogito ergo sum*), was so certain and of such evidence that no ground of doubt, however extravagant, could be alleged by the sceptics capable of shaking it, I concluded that I might, without scruple, accept it as the first principle of the philosophy of which I was in search. (Descartes 1637: IV)

Descartes uses the "doubt" to get to the "truth." More precisely, the "truth" he wants to get to opposes the "opinions" that, as we have seen in the third chapter, correspond to the Greek word δόξα, meaning the "subjective knowledge" that, being particular, contingent, and uncertain, opposes the Greek word ἐπιστήμη, meaning the "objective knowledge" that is universal, absolute and certain. The "truth" that Descartes wants to get to is definitely universal, absolute, and certain: "wholly indubitable," "absolutely necessary," and "so certain and of such evidence." And his solution is the following: what is definitely "true" is thought Thus, thought underpins human existence itself ("I, who thus thought, should be somewhat"). More precisely, it is a matter of a fairly precise kind of thought: "I think, therefore I am (*cogito ergo sum*)." Speaking of the Latin verb *cogitare* means speaking precisely of the kind of thought Descartes shows in his writing, that is, rational thought, as also showed by the etymology of *cogitare*, which means "to shake" (*agitare*) something "within" (*cum*) oneself and, in particular, "shaking" something "within" one's mind. Thus, the modern Western culture is marked by the following crucial idea: rational thought is the way of getting to the "truth" that underpins not only human existence itself but also its autonomy.

We may say that human *logos* increasingly obtain divine prerogatives, starting from the following three: first, the prerogative of getting to the truth; second, the prerogative of underpinning human existence; third, the prerogative of underpinning human autonomy—it is no coincidence that in the modern Western culture and, in particular, in its scientific revolution we can find again the progressive challenge, and even the progressive replacement, characterizing the relationship between the human and the divine. And human *logos* becomes increasingly crucial: it is the human way of both challenging and replacing the divine, as well as of defining the very core of human identity.

The second scientific revolution is another major step of the history of the hypertrophization of *logos*. Ernst Cassirer's words in his *Einstein's Theory of Relativity* may be illuminating: "each measurement contains a purely ideal element; it is not so much with the sensuous instruments of measurement that we measure natural processes as with our own thoughts" (Cassirer 1923: 365), since "the empirical object means nothing but a totality of relations according to law" (Cassirer 1923: 385). In particular, "the general theory of relativity stands methodologically at the end of this series, since it collects all particular systematic principles into the unity of a supreme postulate, in the postulate not of the constancy of things, but of the invariance of certain magnitudes and laws with regard to all transformations of the system of reference" (Cassirer 1923: 404). Indeed,

> What the relativistic physics, which has developed strictly and consistently from a theory of space and time measurement, offers us is in fact only the combination, the reciprocal determination, of the metrical and physical elements. In this sense, we must conceive the pure space-time manifold as the logical *prius*; not as if it existed and were given in some sense outside of and before the empirical

> and physical, but because it constitutes a principle and a fundamental condition of all knowledge of empirical and physical relations. (Cassirer 1923: 442)

Moving from the scientific revolution of the seventeenth century to the scientific revolution of the twentieth century means even radicalizing the role of *logos*: understanding reality is more a matter of using "a purely ideal element," "our own thoughts," "a totality of relations according to law," "theory," "systematic principles," "a supreme postulate," "the system of reference," and "the logical *prius*" than a matter of using "the sensuous instruments of measurements," "the empirical object," and "the constancy of things."

Thus, the privileged human way of getting to "ideal[ity]," "thoughts," "law," "theory," "principles," "postulate," "system," and "logical *prius*," that is, *logos*, becomes even more crucial—we can argue that *logos* becomes hypertrophic: it is what defines the very core of human identity, it is what makes humans able to understand what is, and it is what makes humans able to predict what will be.

Moreover, the works of both Ludwig Wittgenstein, in particular, and the logical empiricists, in general, strongly influenced by Wittgenstein,[1] keep radicalizing the hypertrophization of *logos*. Wittgenstein writes in his *Tractatus Logico-Philosophicus*:

> 3. A logical picture of facts is a thought. 3.01. The totality of true thoughts is a picture of the world. 3.02. A thought contains the possibility of the situation of which it is the thought. What is thinkable is possible too. 3.03. Thought can never be of anything illogical, since, if it were, we should have to think illogically. 3.031. It used to be said that God could create anything except what would be contrary to the laws of logic. The truth is that we could not say what an "illogical" world would look like. 3.032. It is as impossible to represent in language anything that "contradicts logic" as it is in geometry to represent by its co-ordinates a figure that contradicts the laws of space, or to give the co-ordinates of a point that does not exist. (Wittgenstein 1963: 3–3.032)

Interestingly enough, Wittgenstein's argument is represented by the argumentative structure of his *Tractatus Logico-Philosophicus*, that is, seven primary propositions, numbered from one to seven, from which several secondary propositions are articulated. And the isomorphism between argument and argumentative structure, that is, content and form, exemplifies the essential thesis of his *Tractatus Logico-Philosophicus*: there are facts, and there are thoughts, and the latter can represent the former precisely because they share the logical structure—logic and, in particular, logical thought is nothing less than what allows humans to understand reality. Moreover, we can find again a theological issue, which stresses the essential, and even inescapable, role of logical thought: "It used to be said that God could create anything except what would be contrary to the laws of logic." Finally, Wittgenstein's *Tractatus Logico-Philosophicus* ends with the following proposition: "7. What we cannot speak about we must pass over in silence" (Wittgenstein 1963: 7), which means, in particular, that "we cannot speak about" ethics, aesthetics, and metaphysics, since "6.421. It is clear that ethics cannot be put into words. Ethics is transcendental. (Ethics and aesthetics are one and the same)" (Wittgenstein 1963: 6.421), and since "the metaphysical subject [is] the limit of the world—not a part of it" (Wittgenstein 1963: 5.641). Thus,

> 6.53. The correct method in philosophy would really be the following: to say nothing except what can be said, i.e. propositions of natural science—i.e. something that has nothing to do with philosophy—and then, whenever someone else wanted to say something metaphysical, to demonstrate to him that he had failed to give a meaning to certain signs in his propositions. Although it would not be satisfying to the other person—he would not have the feeling that we were teaching him philosophy—this method would be the only strictly correct one. (Wittgenstein 1963: 6.53)

The influence of Wittgenstein's *Tractatus Logico-Philosophicus* on the philosophy of the twentieth century is enormous. Of course, the philosophy of the twentieth century is heterogeneous: it is characterized not only by anti-Wittgensteinian philosophies but also by philosophies inspired by Wittgenstein's later works, which are remarkably different from his *Tractatus Logico-Philosophicus* (see especially Wittgenstein 2009). But its enormous influence, not only on contemporary Western philosophy in particular but also on the contemporary Western culture in general, is surely another major step of the history of the hypertrophization of *logos* and, in particular, of its radicalization.

Yet, the philosopher closest to Wittgenstein, that is, Bertrand Russell, seems to predict the risk of the hypertrophization of *logos* through the following words:

> The concept of "truth" as something dependent upon facts largely outside human control has been one of the ways in which philosophy hitherto has inculcated the necessary element of humility. When this check upon pride is removed, a further step is taken on the road toward a certain kind of madness—the intoxication of power which invaded philosophy [. . .], and to which modern men, whether philosophers or not, are prone. I am persuaded that this intoxication is the greatest danger of our time, and that any philosophy which, however unintentionally, contributes to it is increasing the danger of vast social disaster. (Russell 1946: 856)

Russell's insight is easier to understand if we consider his counterexample: "Locke [. . .] is always sensible, and always willing to sacrifice logic rather than become paradoxical. He enunciates general principles which [. . .] are capable of leading to strange consequences; but whenever the strange consequences seem about to appear, Locke blandly refrains from drawing them. To a logician this is irritating; to a practical man, it is a proof of sound judgment" (Russell 1946: 630). Russell describes the contemporary Western culture ("whether philosophers or not") as follows:

1. We, as Western humans, want to control reality both by understanding what is and by predicting what will be.
2. The tool we have been developing for millennia to control reality is the invention of ideality as its perfect counterpart. More precisely, we have been developing ideality through a progressive focus on logic (to the point that Russell's pupil, that is, Wittgenstein, argues that logical thought allows us to understand reality precisely because the former and the latter share the same logical structure).
3. But, even though logic has been proving for millennia to be our best tool to control reality, its radicalization may be dangerous. More precisely, if logic radically wins out over reality (if the "logician" radically wins out over the "practical man"), then it may lead us to "the greatest danger of our time," "the danger of vast social disaster."
4. In other words, if "the intoxication of power" radically wins out over both "the necessary element of humility" and the "check upon pride," then it may lead us to "a certain kind of madness." As such, we should choose.
5. Our first option is to stop to radicalize logic if we believe that "the intoxication of power" that may lead us to "a certain kind of madness" is more dangerous than surrendering to a reality "largely outside human control" and, therefore, than "sacrific[ing] logic."
6. Our second option is to keep radicalizing logic if we believe the opposite, that is, that our logical control of reality wins out over everything: "the necessary element of humility," the "check upon pride," "sound judgment," and, finally, trying to neutralize "the danger of vast social disaster."

If we go one step further by considering our present, then we should say that Russell's words are actually predictive: not only Western philosophy in particular but also the Western culture in general seem to correspond to his analysis. As for Western philosophy, contemporary analytic philosophy, that is, the leading philosophy of our era, exemplifies a radicalization of logic that may be critical (see at least Baz 2017). As for the Western culture, let us go back to our reading of our technological era.

We may find an instructive analogy between Russell's words and what we are doing when, from an epistemological perspective, we both externalize and meta-externalize from our minds, as well as from our bodies, to algorithmic machines. More precisely, I am trying to argue that both our externalizations and our meta-externalizations are the ultimate result of the phenomenon analyzed by Russell, in that:

1. We, as Western humans, have been increasingly entrusting our control of reality, that is, our understanding of what is and our predictions of what will be, to *logos*, which is a strict form of rationality.
2. We have been increasingly transforming *logos* into an even stricter form of rationality, which is logic—we have been increasingly hypertrophying *logos*.
3. More recently, we have been starting to think, first, that the best form of *logos* is its strictest form and, second, that its strictest form is computation.[2]
4. Finally, we have been starting to believe that the best form of computation is its radical meta-externalization from our minds (considered less powerful) to our machines (considered more powerful) and, in particular, to our algorithmic machines (considered the most powerful), and, therefore, that we should both radically externalize and radically meta-externalize.

The ultimate result is what we have seen in the fourth chapter: we are massively externalizing not only our crucial decisions on health (from what tests we are going to take if we are patients to what tests we are going to prescribe if we are doctors) but also our crucial decisions on everything. And the reason why we are massively externalizing is that we have been starting to think that, paradoxically enough, the best human tool, that is, *logos*, which we have been even hypertrophying, works at its best not when it is human but when it is radically meta-externalized from the human and, in particular, when it is radically automated.

But the question is the following: is it actually intelligent to entrust ourselves to a radically meta-externalized ability of our mind, which is only one of our mind's abilities?

Let us try to answer this question by starting from a possible analogy. When I have to make a difficult decision, I take my time. In particular, I wait at least one day. Other people surely do the same thing: it is no coincidence that the words "sleeping on it" have become idiom. After at least one day, I make my difficult decision using my rational thought, being a philosopher by training. Let us imagine that, today, I have at my disposal an algorithmic machine that can make my difficult decision for me: I have to give it the required data, and it can make my difficult decision for me in less than one second. Thus, I can free myself from being Reichenbach's "eternal Hamlet": "etern[ity]" is reduced to less than one second. But I have another Hamletic doubt: what am I losing when I am shifting my difficult decision from my "sleeping on it" to the algorithmic machine? More precisely, what decision-making processes am I not using when I am shifting my difficult decision from my "sleeping on it" to the algorithmic machine? My Hamletic doubt is that I am not using the decision-making processes that cannot work on data when it comes to the algorithmic machine but that can work on data, even crucial data, when it comes to my mind, since my mind's abilities exceed computability—more precisely, since my mind's abilities exceed rationality, which exceeds *logos*, which exceeds computability. My Hamletic doubt is that I am using only one of my mind's abilities, even though it has been greatly strengthened by automation after having been meta-externalized. On the contrary, even the idiomatic words "sleeping on it" may provide an interesting insight.[3] What if my best decision would partly result from what my mind does when I am sleeping? It is not at all a matter of using only my mind. The algorithmic machine, first, can work on data that is not available to my mind and, second, can work on it in ways that are not available to my mind (for instance, by finding meaningful correlations in less than a second). On the contrary, it is a matter of using both my mind and the algorithmic machine since they may be promisingly complementary.

Again, if you are my doctor, then the most promising scenario is that you and the algorithmic machine work together, since the relationship between your mind and the algorithmic machine may be represented by two circles that only partly overlap. Even though we imagine that the latter is bigger than the former, it cannot work on data we may call ghost data, being data your mind actually has at its disposal even though you totally ignore, first, its existence, second, that your mind actually has it at its disposal, and, third, that your mind actually works on it. It is hard to believe the opposite, that is, that the existence of data that cannot be computable and, therefore, processable by the algorithmic machine, is definitely impossible. Thus, if we want to decrease the risk to lose crucial data and, in particular, crucial ghost data, then the most promising scenario is not to radically entrust ourselves to automation—which means, first, not radically hypertrophying *logos* by assuming that crucial data is reducible to logical data and, second, not radically meta-externalizing *logos* by assuming that crucial data is reducible to data that is computable and, therefore, processable by the algorithmic machine. Thus, if you are my doctor and you happen to understand, somehow intuitively, that something is wrong with me from the look on my face, then the best thing I can do is to take it as not only a piece of information but also as a piece of information that may be crucial, even though it is not computable, logical, and rational.

In both cases, we may speak of abstraction: both our minds and the algorithmic machines select meaningful data from an infinite set of possible data, that is, find meaningful correlations of data from infinite possible correlations of data. But the point is that the sets of possible data from which they abstract do not coincide—the point is that their kinds of abstraction are, again, both different and promisingly complementary.

Let us make one last example. Let us imagine that, when we were children, we had an algorithmic machine at our disposal that could process several sets of data about ourselves, from genetic data to data about our academic performance. And let us imagine that, according to the law of the state in which we lived, any child could have a free university education if their decision on what to study were made by the algorithmic machine that would, in a strictly rational way, identify any child's most promising talent. Otherwise, the child would not have a free university education. Finally, let us imagine that, according to the algorithmic machine, you should have studied mathematics. But, when you started your last year of secondary school, your philosophy teacher was replaced by another one, who changed your life: you not only started to exercise your philosophical abilities but also changed your vision of both the world and yourself. However, at the end of your last year of secondary school, the algorithmic machine kept saying that studying mathematics was the right choice for you (for instance, because data on your past academic performance kept exceeding data on your present academic performance). And your family could not pay for your university. As a result, you studied mathematics and not philosophy. Today, you cannot know how your life would have been if you had studied philosophy and not mathematics, but you can know that you have always regretted not studying philosophy. This example adds another issue to that of the different kinds of abstraction: algorithmic machines are surely conservative, as they surely process past data—and we may ask the following provocative question: are we equally sure that our minds process exclusively past data? If this question is too provocative, we may at least ask the following question: how have we made the best decisions throughout our lives? For instance, when the algorithmic machine said that you should have studied mathematics and not philosophy, what kind of genetic and academic data, and what kind of physical and mental data, was it excluding, since it could not include it in its processing? And, analogously, when the "predictive algorithm" (see the third chapter) says that we should not take the genetic test, what kind of genetic, physical, and mental data is it excluding, since it cannot include it in its processing?

In any case, the human mind is still too mysterious to surrender both to a radical hypertrophization of *logos* and, moreover, to its radical, and even total, meta-externalization from human minds to algorithmic machines. Thus, we should ask ourselves at least the following question: are we actually sure that algorithmic abstraction is always more powerful than mental abstraction? We surely ask ourselves how we can improve the abilities of algorithmic machines (and we surely put a lot of money into it). But do we equally ask ourselves how we can improve the abilities of the human mind? And do we equally put a lot of money into it? It is fairly clear that both our millennial history of progressive hypertrophization of *logos* and our more recent history of its progressive meta-externalization do not help. Indeed, their union even seems to establish a kind of dogma, according to which investigating outside of *logos* is not only unintelligent from an epistemological perspective but also unworthy from an ethical perspective.

However, if we go back to Plato's words, according to which "explanation [. . .] [is] the interpretation of your difference," then is it actually intelligent, and is it actually worthy, to radically entrust yourself to automation, which, by definition, brings down heterogeneities, that is, differences, and brings up homogeneities? What if you are the difference? In Reichenbach's words, "shall I be right in *this* case?" (see at least Noble 2018; O'Neil 2016). What happens to the exceptions of your present? And what happens to the surprises of your future? Again, is it actually intelligent, and is it actually worthy, to radically entrust yourself to what, by definition, has limited abilities when it comes to valuing differences (i.e., algorithmic machines) and to radically atrophy what has mysterious abilities when it comes to valuing differences (i.e., your mind)?

If we move from philosophical questions to a possible literary answer, then, as we have seen in the first chapter, McEwan's Turing says something interesting:

> Solving maths problems is the tiniest fraction of what human intelligence does. We learned from a new angle just how wondrous a thing the brain is. A one-litre, liquid-cooled, three-dimensional computer. Unbelievable processing power, unbelievably compressed, unbelievable energy efficiency, no overheating. But so far science has had nothing but trouble understanding the mind. So, knowing not much about the mind, you want to embody an artificial one in social life. Machine learning can only take you so far. You'll need to give this mind some rules to live by. How about a prohibition against lying? According to the *Old Testament*, *Proverbs*, I think, it's an abomination to God. But social life teems with harmless or even helpful untruths. How do you separate them out? Who's going to write the algorithm for the little white lie that spares the blushes of a friend?

If the objection is that McEwan's Turing is nothing more than a character in a novel, then we can move from literature to the Horizon 2020 program, which is officially defined as "the biggest EU research and innovation programme ever with nearly €80 billion of funding available over 7 years."[4] If we read the webpage on its program, then we find the following sections:[5]

1. "Excellent Science," including two out of four subsections on technology: "Future and Emerging Technologies" and "Research Infrastructures, Including E-infrastructures".
2. "Industrial Leadership," including three out of three subsections on technology: "Leadership in Enabling and Industrial Technologies," "Access to Risk Finance," and "Innovation in SMEs".
3. "Societal Challenges," including six out of seven subsections at least partly on technology, three of which on technology: "Secure, Clean and Efficient Energy," "Smart, Green and Integrated Transport," and "Climate Action, Environment, Resource Efficiency and Raw Materials".
4. "Science with and for Society," including six out of six subsections at least partly on technology.
5. "Spreading Excellence and Widening Participation," focusing on institutional infrastructure.
6. "Enhanced European Innovation Council (EIC) Pilot," including four out of four subsections on technology: "EIC Accelerator Pilot," "Fast Track to Innovation," "Future and Emerging Technologies (FET) Open," and "Horizon Prizes".
7. "Focus Areas," focusing on four out of four technological areas.
8. "European Institute of Innovation and Technology (EIT)," focusing on institutional infrastructure.
9. "Euratom," including one out of one subsection on technology: "Nuclear Fusion".

Moreover, if we read the webpage on the areas of the Horizon 2020 program, then we find twenty-four sections:[6] twenty sections are at least partly focused on technology, two sections are focused on institutional infrastructures ("Funding Researchers" and "International Cooperation"), and two sections are focused on social sciences and humanities ("Social Sciences and Humanities" and "Society").

It goes without saying that the great focus on technology is both intelligent and worthy, since, as we have learned from the historical evolution of the figure of Prometheus, humans are essentially technological. But the question is the following: where is the human mind? The answer is that, even though it is sometimes possible to find something on it in scientific research and philosophical research, it is far from being considered crucial. Indeed, if you search for the words "human mind" in the Horizon 2020 website, then you find only six items, two of which are not on research and six of which are on the "Human Brain Project (HBP),"[7] which is focused not on the human mind but, meaningfully, on the human brain, that is, not on what, according to the Oxford English Dictionary, is "the seat of awareness, thought, volition, feeling, and memory,"[8] but on what, according to the Oxford English Dictionary, is "the organ contained in the skull of humans and other vertebrates."[9] It goes without saying that both these issues are crucial for humans. But it equally goes without saying that humans and, in particular, Western humans are infinitely better trained to investigate the latter than the former—and the reason why Western humans are infinitely better trained to investigate the human brain than the human mind is that the former is an infinitely better fit for the tool on which Western humans have been training themselves to rely: again, *logos*, which has been, first, increasingly hypertrophied and, second, increasingly meta-externalized. *Logos* and, in particular, what we may call technological *logos* is surely an outstandingly successful tool, which allows us to remove a brain tumor, for instance. But what happens to "awareness, thought, volition, feeling, and memory"? What happens to the human mind? In the second and fourth chapters, I have said that I was pleased to stop to continuously train my memory using technologies, from handwriting to digital writing. But I can honestly say that I am not equally pleased to stop to continuously train my "awareness, thought, volition, feeling," since I identify not only my actual identity but also the kind of identity I want to have as a human being, as well as the kind of life plan I want to have as a human being, precisely with my "awareness, thought, volition, feeling," even though, paradoxically enough, I am quite ignorant of what they are precisely, of what my mind is precisely.

However, this paradox is illuminating, since it means that, even though we are quite ignorant of both what kinds of data our mind can access and what kinds of data processing our mind is capable of, the following first and second options seem to win out over the third option:

1. We decide that, even though we are quite ignorant of the actual powers of our mind, we radically meta-externalize its activities and, in particular, the activities coinciding with its strictest form of rationality, because we think that the strictest form of rationality is not only the best tool in general but also the best tool strengthened by automation in particular.
2. We decide that, even though we are quite ignorant of the actual powers of our mind, we radically meta-externalize its activities and, in particular, the activities coinciding with its strictest form of rationality, because we are too lazy: first, we think that studying the actual powers of our mind is too hard and, second, we think that a smart way of killing two birds with

one stone is the radical meta-externalization of what we consider its best tool, since, through its meta-externalization, we can more easily study and use its actual powers.[10]

3. We decide that, precisely because we are quite ignorant of the actual powers of our mind, we think that a smart way of killing two birds with one stone is the following: on the one hand, by not radically meta-externalizing what we consider one of its best tools, we can more easily study and use its actual powers without radically atrophying their human use and, on the other hand, by not radically atrophying their human use, we can invest their human use in studying the actual powers of our mind, from its strictest form of rationality to what is not a form of rationality at all and, therefore, is still too mysterious.

Of course, as we have stressed, studying the human mind without being afraid of challenging not only its complexity but also what we have called a kind of (Western) dogma (according to which investigating outside of *logos* is both unintelligent from an epistemological perspective and unworthy from an ethical perspective) requires two of the hardest things to obtain: courage and money. But, since money is, again, a matter of courage (at least in the Western states), we can say, in Reichenbach's words, that it is a matter of deciding where "is the point where [. . .] [we] lack the courage." We are paradoxically externalizing almost everything because we "lack the courage" to do what? According to what we have reflected upon, the most precise answer, at least from a philosophical perspective, seems to be the following: we "lack the courage" to face a challenge so complex that we are likely to fail—more precisely, we "lack the courage" to face a challenge so complex that, by making us fail, is likely to make us fail what seems to be our actual challenge: moving from an imperfect ontological status (whose imperfection implies effort, not allowing our laziness) to a perfect ontological status (whose perfection does not imply effort, allowing our laziness). Again, it is a matter of moving from a human ontology to a divine ontology even by technologically creating it, since its total immanence may guarantee that it may be what our epistemological ambition, together with our ethical laziness, seems to desire: a kind of divine servant of ours, allowing us to definitely free us from everything (if nothing goes wrong).

NOTES

1. Strongly influenced by Russell (see at least Russell and Whitehead 1963).
2. As it is standardly defined, for instance, in the Cambridge English Dictionary: "the act or process of calculating an answer or amount using a machine" (quote from

https://dictionary.cambridge.org/it/dizionario/inglese/computation). Thus, computation relies on something precise, that is, quantification (which may be not always possible).

3. For instance, work concerning the relationship between the emotional sphere and the epistemic sphere is exponentially increasing (see at least Brady 2013).

4. Quote from https://ec.europa.eu/programmes/horizon2020/what-horizon-2020.

5. See https://ec.europa.eu/programmes/horizon2020/h2020-sections.

6. See https://ec.europa.eu/programmes/horizon2020/find-your-area.

7. Quote from https://ec.europa.eu/programmes/horizon2020/en/search/site/human%20mind.

8. Quote from https://www.oed.com/view/Entry/118732?rskey=SASQI8&result=1&isAdvanced=false#eid.

9. Quote from https://www.oed.com/view/Entry/22536?rskey=dfawmW&result=1&isAdvanced=false#eid.

10. See especially artificial neural networks (see at least Amit 1989).

Chapter 5

Farewell to Humans?

A POSSIBLE SCENARIO

Thanks to an anonymous reviewer of a publication of mine (see Chiodo 2020), I realized that a promising way of addressing, at least, the last issue of the fourth chapter may be trying to imagine a possible scenario that takes it to its extremes.[1] Thus, let us try to provocatively use one of the best results of imagination, that is, literature and, specifically, the historical evolution of the figure of Prometheus. And let us try to imagine the following possible dialogue between Prometheus and Zeus happening less than a century from now:

"Prometheus, what are you looking at with such astonishment on your face?"

"I am looking at humans if they can still be considered humans at all."

"Why would you say such a thing? You happened to be the one who qualified them as humans: neither as strong as tigers nor as fast as cheetahs but even better qualified, thanks to your love for them. Indeed, almost as good as Hephaestus at mastering the divine fire and almost as good as Athena at mastering the divine wisdom in the arts. Humans are surely the wisest of the living creatures!"

"They should be the wisest of the living creatures, but they are not."

"How can you say such a thing? Can you not see how good they are at mastering your gifts? They became even better than Hephaestus and Athena themselves at mastering fire and wisdom in the arts! They succeeded in finding remedies to the evils of Pandora's jar! Look at how they could not only relieve physical pain with anesthetics, baneful anxieties with anxiolytics, and wretched diseases with surgeries but also remove them definitively through the development of sophisticated bioengineering technologies that can now entirely bear the burdens of those physical pain, baneful anxieties, and wretched diseases that, only decades ago, used to afflict humans, sometimes causing their deaths. And look at how they could not only relieve harsh labor with machines but also remove it definitively through the development of sophisticated mechanical engineering technologies that can now entirely bear the burdens of that harsh labor that, only decades ago, used to afflict humans, sometimes causing their deaths. And look at how they could not only relieve disasters with weather forecasts but also remove them definitively through the development of sophisticated data science technologies that can now entirely bear the burdens of those disasters that, only decades ago, used to afflict humans, sometimes causing their deaths."

"But can you not see what they have turned into?"

"Yes! They have turned into us, living lives without pain, anxieties, diseases, labor, and disasters. They have turned into potentially everlasting beings. Prometheus, you happened to be the one who loved humans most. Are you not happy to see that, finally, they can equal us? Do not tell me that you are envious!"

"I still happen to be the one who loves humans most! That is the reason why you see such astonishment on my face!"

"I do not follow you."

"Can you not see what they have done? Look at that human over there, who is asking his algorithmic machine what to do with himself. I have been watching him for three days. He has been asking his algorithmic machine everything, from what to eat to what to think. He was listening to a political debate online about whether the possibility of living an everlasting life should be given to all humans or just a select few, and he even asked his algorithmic machine what he should have thought!"

"But this is wise! His algorithmic machine can make it easier for him to think!"

"Not at all! His algorithmic machine is making him stop to think! Can you not see? I gave humans the best gifts, even divine gifts. I even stole to give them Hephaestus' fire and Athena's wisdom in the arts. But look at what they are doing! They are depriving themselves of my gifts to give them to their algorithmic machines! This is absurd!"

"But . . ."

"There is no but! Can you not see what it means to ask an algorithmic machine everything from what to eat to what to think? I can see what it means! It means that all the good I have done for humans is useless! Do you not remember what Aeschylus said? 'I made them have sense and endowed them with reason.' And now they are giving their sense and reason away! And when they 'dwelt beneath the ground like swarming ants,' 'though they had eyes to see, they saw to no avail; they had ears, but they did not understand.' And now they are giving their avail and understanding away! And when they 'dwelt beneath the ground like swarming ants,' they were 'just as shapes in dreams,' and 'without purpose they wrought all things in confusion,' and 'managed everything without judgment.' And now they are going back to be 'just as shapes in dreams,' 'without purpose,' 'in confusion,' and 'without judgement!' Look again at that human over there, who is no longer a human, not as I qualified them. Can you not see what I see? I see nothing more than the kind of 'swarming ants' I saved humans from being. Aeschylus said that humans were 'like swarming ants, in sunless caves.' Look again at that human over there, or at what is left of him as a human. Since I started to watch him three days ago, he has been living in his 'sunless cave' the whole time. He 'had no sign either of winter or of flowery spring or of fruitful summer,' since all he has done has been done by his algorithmic machine. He did not take a step out of his 'sunless cave.' As you can see, I have failed. Finally, humans have turned into 'swarming ants.' What a waste of divine gifts! I wanted them to become even better than Hephaestus and Athena themselves at mastering fire and wisdom in the arts, but they have given them away. I wanted them to be the best of the living creatures, but they have become the worst: not even able to decide what to eat. Zeus, can you not see what they have turned into? They have turned into the most defenseless of the living creatures!"

"Prometheus, do not despair. They can rely on what they can create with your gifts: not only their algorithmic machines that can make all their decisions but also the other machines that can remove all their physical pain, baneful anxieties, wretched diseases, harsh labor, and disasters. I see evolution where you see involution! Did you not give them your gifts to make them the most defenseless and, at the same time, the most powerful of the living creatures? Did you not give them your gifts to make them the special earthly beings who can create their own servants that can free them from any burden?"

"I gave them my gifts to make them even more powerful! Can you not see that my gifts are enormously more powerful than the machines they create? This is absurd! They have gold in their own minds but use something made of silicon. They have Hephaestus' and Athena's weapons but use wooden sticks. What still undiscovered powers they have in their own minds! But they waste their time playing the easiest game: looking at what there is out and not at what there is in. What a waste!"

"But they may want to look at what there is out! It may be wise!"

"Wise? Is it wise to give what is most powerful away? Is it wise to give their own human nature away? Humans! I wanted them to be the most privileged beings: superior to the other earthly beings and inferior to the gods in a special way, which makes them even better than the gods. Can you not remember what Ovid said? The other earthly beings 'look on all fours at the ground.' Even we, as gods, look 'at the ground.' I have been watching that human over there for three days. And, when I do not look 'at the ground,' I look at you, right in front of my eyes. But can you not remember what Ovid said about humans? He said that I gave them 'an upturned face, and told them to lift their eyes to the stars.' Can you not see what it means? Humans are absolutely the most privileged beings, even superior to the gods! They can have infinite challenges 'to lift their eyes to,' whereas we cannot! They can have infinite evolutions 'to lift their eyes to,' whereas we cannot!"

"But . . ."

"There is no but!"

"There is a but! Listen to me. Prometheus, can you not see that there is still something that is precisely as you say? Continuously challenging, even when humans stop. Continuously evolving, even when humans stop."

"Zeus, I do not understand your point."

"Prometheus, my point is simple: their machines! Their machines! Can you not see that their machines are becoming precisely who humans used to be? Look at the sophisticated bioengineering technologies that now bear the burdens of physical pain, baneful anxieties, and wretched diseases: while humans are totally forgetting what they are, their machines are learning precisely what human pain, human anxieties, and human diseases are, even more so than humans themselves. And look at the sophisticated mechanical engineering technologies that now bear the burdens of harsh labor: while humans are totally forgetting what it is, their machines are learning precisely what human labor is, even more so than humans themselves. And look at the sophisticated data science technologies that now bear the burdens of disasters: while humans are totally forgetting what they are, their machines are learning precisely what human disasters are, even more so than humans themselves. Moreover, look at the sophisticated algorithmic machine of that human over there: while that human over there is totally forgetting what it means to make a decision, from what to eat to what to think, his algorithmic machine is learning precisely what it means to make a human decision, from the easiest to the hardest, even more so than that human over there himself."

"Are you saying that the machines are becoming human, in that they are obtaining all the qualities that used to be human?"

"It makes sense."

"And are you also saying that humans are becoming mechanical, in that they are obtaining all the qualities that used to be mechanical?"

"It makes sense, too."

"Zeus, but I am afraid that I can see nothing good where you can. If I think of humans, the most privileged beings, as mechanical beings, then my despair becomes even deeper! To what end have I sacrificed even myself? To make the most animate of the living creatures become the most inanimate? To make them less animate not only than the other living creatures, and even than the gods, but also than the machines? Zeus, I have failed, and failed miserably."

"Prometheus, I understand your point, but, please, try to understand mine. You are right to say that you can see nothing good where I can. You say that you have sacrificed even yourself to give humans their special qualities. But what if humans are doing to the machines precisely what you have done to them? What if they are scarifying even themselves to give the machines their special qualities?"

"Zeus, I am even more astonished now."

"Prometheus, please, try to keep listening to me. There may be a special reason why you did it. And there may be a special reason why humans are doing it. Moreover, your reasons may be analogous. What was your reason?"

"I do not know what to say. I can only say that I realized that, when my brother Epimetheus was asked by the Olympians to assign a given number of good qualities to the living creatures, he totally neglected humans, and I felt pity. Yes, I felt pity for humans."

"And pity may be what humans are feeling."

"Pity? Pity for the machines?"

"Not necessarily. It may be pity for themselves."

"What? Pity for the most privileged beings?"

"Prometheus, can you not see that the burden of being the most privileged beings may be the most unbearable burden? Ovid said that you gave them 'an upturned face, and told them to lift their eyes to the stars.' Can you not see that being driven to infinite challenges and evolutions may be the most unbearable burden?"

"Are you saying that humans are trying to give away not my gifts, which are sense, reason, avail, understanding, and the ability to master fire and wisdom in the arts, but the burdens they bring, which are the most unbearable burdens?"

"Precisely."

"What a paradox! It seems that I have done the worst thing to those I have always loved the most!"

"A classic puzzle: great love may mean great danger."

"Even the danger of saying farewell to humans?"

"It may be. But it may be also the birth of someone else."

"What do you mean?"

"I mean the birth of new living creatures: machines that will actually become human, in that they will actually have sense, reason, avail, understanding, and the ability to master fire and wisdom in the arts."

"But they will never have the human mind as I qualified it, which can make humans able to face the most unpredictable exceptions of the present, and even the most unpredictable surprises of the future! And it can even allow them to be the most unpredictable of the living creatures!"

"You may be right. But you know that the truth is that, if humans stop to investigate, and even using, the human mind as you qualified it, then we cannot know its true powers. You and I have created and qualified the human mind as the most powerful thing ever precisely by making it the most unpredictable thing ever. But, if it stops to work, then its still mysterious powers will never be known. And we will never know whether there is actually something that categorically distinguishes humans from machines or not."

"Zeus, can you not see that I was right when I said that I have failed miserably?"

"Prometheus, what I can see is that we cannot know whether or not you were right. Humans may be doing the worst thing ever or the best thing ever. They may be giving the most powerful thing away. Or they may be creating the machines as the most powerful thing. I can agree with you that the wisest thing to do may have been to keep investigating the human mind's mysterious powers before replacing them with the machines. But, my dear Prometheus, can we actually blame them for not wanting to bear the most unbearable burden of being driven to infinite challenges and evolutions?"

"You may be right, my dear Zeus. Yet, my despair becomes ever deeper."

A POSSIBLE REFLECTION UPON THE POSSIBLE SCENARIO

Of course, the meaning of imagining a possible dialogue between Prometheus and Zeus is only a provocative way of taking our reasoning to extremes, since taking something to extremes may be a promising way of reflecting upon exceedingly complex issues. And it goes without saying that the meaning of the possible dialogue I imagined is not trying to predict anything at all. Of course, any kind of prediction may be totally wrong. But trying to imagine a possible scenario may be a promising way of doing what Shakespeare says in his *Hamlet*, that is, being ready, since "the readiness is all" (Shakesp. *Haml.* 3954–5).

Thus, let us try to reason on at least three issues:

1. The relationship between the imagination of a possible scenario that takes our reasoning to extremes and what we have seen in the fourth chapter: atrophying our abilities by meta-externalizing everything from the human to the technological is likely to mean becoming as powerless as the pre-technological ant-like humans described by Aeschylus, that is,

paradoxically enough, de-evolving into a pre-technological ontological status.

2. The relationship between radical meta-externalization and identification of the very core of human identity, especially if, paradoxically enough, what is radically meta-externalized is precisely what has been thought of as the very core of human identity, that is, exercising *logos* (at least in the Western culture).
3. The relationship between bearing an unbearable burden and anarchism.

As for the first issue, it may seem nonsensical that the more we make our technologies evolve, the more we risk to de-evolve into a pre-technological status. However, let us look at a simple example that may provide us with another perspective (it is a simple example, but it is true: it really happened to me). Two days ago, at noon, I happened to check my smartphone and realized that it was not working. I turned it on and off again and again, but nothing worked. I realized that I could not do at least two important things: first, remembering what I had to do in the afternoon; second, registering my students' marks, which needed to be registered immediately (our online register has to be signed by using a one-time password sent via text to our smartphone).

Surprisingly enough, I started to feel as defenseless as Aeschylus' "swarming ants" and Plato's "naked, unshod, unbedded, unarmed" humans—surprisingly enough, I started to panic. I also realized that, if I could not find a solution by that night, then I would have other reasons to feel radically defenseless: first, I would not be able to receive at least two important calls, the first private and the second professional; second, I would not be able to remember what I had to do in the coming days. Fortunately, I managed to find a quick fix thanks to a colleague of mine. But I learned that it is not nonsensical that atrophying our abilities by meta-externalizing everything from the human to the technological is likely to mean becoming as powerless as the pre-technological ant-like humans described by Aeschylus, that is, paradoxically enough, de-evolving into a pre-technological ontological status.

We may easily imagine to extend my simple example to a bigger scenario and ask ourselves, for instance: what if a hacker took down the operating system of a national government? And what if it happened in the middle of a severe national emergency?[2] Even though we do not consider extreme scenarios, we can learn the following:

1. Indeed, giving content away may mean giving abilities away, as we have seen in the fourth chapter—again, externalizing may mean meta-externalizing, and meta-externalizing is precisely what may make us as defenseless as Aeschylus' "swarming ants" and Plato's "naked, unshod, unbedded,

unarmed" humans. What I learned from my simple example is that I was defenseless not only in terms of my memory (which, as I have said, I am pleased to stop to continuously train) but also in terms of my control of reality, from what is going on outside of me (my schedule, my students, my private relationships, my professional relationships, etc.) to what is going on inside of me, which is even more important, since it means my control of my physical, emotional, and intellectual reactions, as well as of my physical, emotional, and intellectual resources. More precisely, I learned that realizing that I am defenseless means, at least at first, panicking, and panicking means losing control, and, therefore, using physical, emotional, and intellectual resources badly, leading to committing mistakes.

2. Indeed, giving abilities away, that is, meta-externalizing, may make us de-evolve into a pre-technological ontological status, in that what may happen is the following kind of paradox: as we have seen, we are essentially technological, but, if we radically lose control of the relationship between ourselves and our technology, then we radically lose our technological power. Figuratively, let us imagine that we have two vessels. If both vessels keep holding water, then, when vessel A breaks away from vessel B for any reason, the latter will keep holding water anyway. On the contrary, if vessel B does not keep holding water, then, when vessel A breaks away from vessel B for any reason, the latter will not keep holding water anyway.

Let us translate the figurative case into the literal case: water represents technological power, that is, awareness and, therefore, control of technology. If we keep having technological power, then, when technology gets out of our control for any reason, we will remain essentially technological, since we will retain the ability to manage, and even create, technology (and I will be able, for instance, to fix my smartphone myself and, therefore, to continue with my day). On the contrary, if we do not keep having technological power, then, when technology gets out of our control for any reason, we will not remain essentially technological, since we will not retain the ability to manage, and even create, technology (and I will not be able, for instance, to fix my smartphone myself and, therefore, to continue with my day: I will not be able to remember, to control my schedule, students, private, and professional relationships, and to control my physical, emotional, and intellectual reactions, and resources. I may summarize by saying that I will be paralyzed—and being paralyzed means not keeping being technological, that is, able to be what technology means, as we have seen in the first chapter: a survivor and a ruler).

If it is true that meta-externalizing may make us de-evolve into a pre-technological ontological status by making us stop to be survivors and rulers, then it is also true that the ultimate result of meta-externalizing is analogous to the severest meaning of being "vulnerable," which, according to the Oxford English Dictionary, literally means something "that may be wounded; susceptible of receiving wounds or physical injury,"[3] and figuratively means something "open to attack or injury of a nonphysical nature."[4] The analogy between being vulnerable and stop to be survivors and rulers leads us to think that if we do not want to be "open to attack" both in "physical" terms and in "nonphysical" terms, then we should do one of the following two things: on the one hand, we can decide not to radically meta-externalize (and even to legislate accordingly) and, on the other hand, we can decide to radically meta-externalize anyway (what if the possible dialogue between Prometheus and Zeus is somehow right to imagine that what humans actually want is to stop to bear the most unbearable burden of being precisely who they have been for millennia in the Western culture?). In any case, we should make a decision. And, if we want to keep meta-externalizing, even radically, but without ending up being totally "open to attack," then we should work on at least one thing, from an epistemological perspective: trying to move what we do from irremediability to remediability (I have worked quite extensively on this issue, especially in Chiodo 2011).

As for the second issue (the relationship between radical meta-externalization and identification of the very core of human identity), the logical paradox is the following: first, we, as Western humans, have been founding the very core of our identity on the exercise of *logos* itself; second, we have been increasingly radicalizing the relationship between the former and the latter; third, we have been increasingly meta-externalizing the latter, especially in our technological era; fourth, what will constitute the very core of our identity, especially in the future? The logical paradox is clear in the last two steps: we are radically meta-externalizing precisely what we have been defining for millennia as the very core of the special earthly beings we are—we are radically meta-externalizing precisely what we have been defining for millennia as the most essential distinction between humans and other living creatures.

The first philosophical question is why we are doing such a paradoxical thing. Again, a promising answer can be given by the historical trajectory that, fairly precisely, and most instructively, moves (in the Western culture):

1. From the crucial invention of ideality as the perfect counterpart of reality, as well as from the crucial decision that the best tool to use the former to perfect the latter is *logos*, through its abilities to abstract and idealize
2. To the first phase of the crisis of ideality, which means trying to make the ideal real—we can say, now, that it ultimately means trying to bear the

most unbearable burden, that is, to be perfect, as well as to use technology to succeed
3. To the second phase of the crisis of ideality, which means trying to abolish the ideal—we can say, now, that it ultimately means trying to give the most unbearable burden away
4. Finally, to our technological era, which may be read as the most radical form of anarchism we have ever experienced precisely because anarchism may be the best way of trying to give the most unbearable burden away.

We get to the third issue, which is the relationship between bearing an unbearable burden and anarchism. I shall also try to address this issue in the sixth chapter, but let us start from its essential meaning. We have seen that, through technology, we are radically meta-externalizing precisely what we have been defining for millennia as the very core of our identity as Western humans. We can now add that radically meta-externalizing the very core of our identity even strengthens anarchism, in that it means a kind of double abolition of ideality. On the one hand, as we have seen in the second chapter, it is a matter of abolishing ideality as something external to us, from an ideal as a model (meaning ideality in a strict sense) to an expert as a mediator between us and the ideal model (meaning ideality in a broad sense). On the other hand, we can now add that it is also a matter of abolishing ideality as something internal to us, that is, internalized precisely in the sense meant by Kant through his words according to which "we have in us no other standard for our actions than the conduct of this divine human being, with which we can compare ourselves, judging ourselves and thereby improving ourselves, even though we can never reach the standard." Kant's "divine human being" that "we have in us" is a condition of possibility for another essential notion we have seen in the second chapter, that is, human autonomy: any human being can be "autonomous" if the "law" they follow is "self-given." Kant teaches us that humans can be both heteronomous and autonomous and, moreover, that the latter condition is the evolution of the former condition, especially from an ethical perspective. Autonomy as an evolution of heteronomy is a classical issue in the modern Western culture: we, as Western humans, follow laws given by something outside of us and, moreover, by something inside of us. And Kant's masterful lesson adds something even more essential: when we follow laws given by something inside of us, we use our reason, as Kant defines it[5]—when we follow laws given by something inside of us, we use what is closest to the very core of our identity as Western humans.

Thus, the relationship between bearing an unbearable burden and anarchism becomes clearer: when we doubly abolish ideality and, in particular, when we abolish ideality as something internalized, we totally abolish any kind of "ruler"—we totally move to "rulerlessness," that is, the very definition of "anarchism." First, we dissolve external "rulers," such as ideal models and experts. Second, we dissolve internal "rulers," starting from what Kant calls reason and autonomy. More precisely, we dissolve internal "rulers" by shifting our reason and autonomy from ourselves to technology. Kant's reason and autonomy are what can answer the crucial question on the condition of possibility for our decisions and actions, starting from the following: "ask yourself whether, if the action you propose were to take place by a law of the nature of which you were yourself a part, you could indeed regard it as possible through your will" (Kant 1788: 5, 69). In particular, "one says: if everyone permitted himself to deceive when he believed it to be to his advantage, or considered himself authorized to shorten his life as soon as he was thoroughly weary of it, or looked with complete indifference on the need of others, and if you belonged to such an order of things, would you be in it with the assent of your will?" (Kant 1788: 5, 69)

Kant's words become even touching in our technological era, if we think of the algorithms that actually decide what our "advantage" is, what our "life" expectation is, and what "others" think of us. Moreover, Kant's words become even more touching: when we read "to shorten his life as soon as he was thoroughly weary of it," we can find precisely what, from Plato to Descartes to Wittgenstein, as well as from Sophocles to Shakespeare, from Pirandello to Reichenbach, has been one of the essential questions of the Western culture: how can we bear the burden of being "weary" of our lives?

Again, our ultimate answer seems to be that we can bear the burden of being "weary" of our lives by becoming radically anarchistic—more precisely, our ultimate answer seems to be that we cannot bear the burden and, therefore, we give away every kind of ruler, especially if it is the most unbearable: our internal ruler, which means our burden of being driven to infinite challenges and evolutions. And the kind of technology we are creating can be precisely our way of becoming radically anarchistic: technology can finally bear the burden of being "weary" of our lives—technology can finally bear the burden of what has been for millennia the very core of our identity as Western humans.

NOTES

1. I could not imagine a possible scenario in that publication of mine, being a journal article with a necessarily limited length, but I promised to try to do it in the future. This book may be the right opportunity.
2. While I am writing, we are facing the coronavirus emergency.

3. Quote from https://www.oed.com/view/Entry/224872?redirectedFrom=vulnerable#eid.

4. Quote from https://www.oed.com/view/Entry/224872?redirectedFrom=vulnerable#eid.

5. See especially the paragraph *On the primacy of pure practical reason and its connection with speculative reason* (Kant 1788: 5, 199–221). See also Kant 1788: 5, 33: "Autonomy of the will is the sole principle of all moral laws and of duties in keeping with them; heteronomy of choice, on the other hand, not only does not ground any obligation at all but is instead opposed to the principle of obligation and to the morality of the will." See, finally, Kant 1788: 5, 15: "[In the practical use of reason], reason is concerned with the determining grounds of the will, which is a faculty either of producing objects corresponding to representations or of determining itself to effect such objects (whether the physical power is sufficient or not), that is, of determining its causality. For, in that, reason can at least suffice to determine the will and always has objective reality insofar as volition alone is at issue."

Chapter 6

The Most Radical Form of Anarchism We Have Ever Experienced

PAST AND PRESENT FORMS OF ANARCHISM

As we have seen in the second chapter, anarchism as an epistemological way of facing human life has systematically characterized the Western culture since the twentieth century (see at least Fiala 2017; Franks and Wilson 2010; Jun 2017; Jun and Wahl 2010; Marshall 2010; McLaughlin 2007), even though there were references to anarchism in particular, as well as to anarchy in general, as early as the nineteenth century (for instance, by Bakunin's words published in 1882: "if God really existed, it would be necessary to abolish him," Bakunin 1970: 28).[1] In 1907, James describes his philosophy by starting to use the word "anarchistic" from an epistemological perspective: "a radical pragmatist is a happy-go-lucky anarchistic sort of creature" (James 1981: 116). And, in addition to the philosophers we considered in the second chapter, French philosophers have also approached anarchism from an epistemological perspective, from Deleuze (see at least Deleuze 1983) to Foucault (see at least Foucault 1973) to Derrida (see at least Cornell, Rosenfeld and Carlson 1992). The epistemological anarchism we are focusing on has also ethical and political consequences, as we have seen in the second chapter (see at least Nozick 1974, proposing the notion of minimal state against the notion of individual anarchism).

As the Oxford English Dictionary reports, the articulations of the notion of anarchy in general, which sometimes falls into its radicalization, that is, the notion of anarchism in particular, are several: "1.a. Absence of government; a state of lawlessness due to the absence or inefficiency of the supreme power; political disorder. [. . .] b. A theoretical social state in which there is no governing person or body of persons, but each individual has absolute liberty (without implication of disorder). [. . .] 2.a. [. . .] Absence or nonrecognition of authority and order in any sphere. [. . .] b. Nonrecognition of moral law; moral disorder. [. . .] c. Unsettledness or conflict of opinion."[2] Even though our focus is on the notion of anarchism, and not of anarchy, and, more precisely, on the notion of epistemological anarchism (together with its several consequences on the general way of facing human life, starting from ethical and political perspectives), it may be useful to address the meanings of anarchy as reported by the Oxford English Dictionary. Indeed, the comparison between its definitions and our reflections on anarchism can further clarify the meanings of the latter through the meanings of the former. I think that one issue is particularly illuminating: moving from anarchy (which has somehow characterized the Western culture since its genesis)[3] to its radicalization (which characterizes the contemporary Western culture), that is, moving from anarchy to anarchism, means reforming the notion of autonomy from a theoretical perspective.

As we have seen in the third chapter, as well as in the fifth chapter, the notion of "autonomy" is founded, even etymologically, on the notion of "law," since being "autonomous" means following a "law": the "law" that is "self-given." The meanings of anarchy reported by the Oxford English Dictionary tell us that, if there is anarchy, then there is the absence of something specific and, at the same time, there is the presence of something equally specific:

1. What is absent is "government," "supreme power," "governing person or body of persons," "authority," "order," and "moral law."
2. What is present is "state of lawlessness," "absolute liberty," and "unsettledness or conflict of opinion." There may be "disorder," such as in the cases of "political disorder" and of "moral disorder," or there may not be "disorder," such as in the case of "without implication of disorder."

We can translate what is absent and what is present into the following terms:

1. The *arché* coinciding with the heteronomous "law" is absent.
2. The *arché* coinciding with the autonomous "law" is present.

More precisely, what is present is the kind of "absolute liberty" that can be described by Godwin's words quoted in the footnote: "Above all we should not forget that government is an evil, a usurpation upon the private judgment and individual conscience of mankind." Thus, speaking of anarchy means speaking of an "absolute liberty" that is founded on the presence of "private judgment" and "individual conscience"—speaking of anarchy means speaking of an "absolute liberty" that is founded on autonomy in a strictly etymological sense, according to which the *arché* coinciding with the autonomous law gives everyone "private judgment" and "individual conscience."

What if we move from anarchy to anarchism? The answer is that what changes between the former and the latter is precisely the notion of autonomy, being reformed toward a progressive dissolution.

Let us try to develop the argument of the fifth chapter according to which the form of anarchism characterizing our technological era means the dissolution of internal rulers (we may say internal laws: the internal *arché*, in any case) by shifting our autonomy from ourselves to technology. Let us look at an example. One week ago, an Italian Academy of Fine Arts invited me to a roundtable about a critical situation it was facing (together with other European Academies of Fine Arts):[4] the most popular social networks blocked the advertisements for the art exhibition "Tiziano e Caravaggio in Peterzano" organized by the Academy of Fine Arts because of the nudes in the paintings. Of course, it was not a case of (ethical) censorship but of (technological) block. The algorithms of the most popular social networks are trained, first, to recognize and, second, to block all nudes (even if they are the greatest masterpieces in the history of art). There are several clear advantages. But the Academies of Fine Arts had to face a critical situation: even though the social networks finally could find a solution, the Academies of Fine Arts could not advertise their art exhibitions for several weeks. Thus, our question is the following: what happens to our "absolute liberty" and, in particular, to our "private judgment" and "individual conscience," when we rely totally on algorithms, which may be inscrutable black boxes, and not on human beings? More precisely, what happens to our "private judgment" and "individual conscience" when, as we have seen in the third chapter, we rely totally on something more opaque and rigid, we cannot negotiate with, and not on someone less opaque and rigid, we can negotiate with? If the decision-maker is a human being, from a "governing person" to a "body of persons," or even a "government" coinciding with a "supreme power," then two things may happen: first, there is no critical situation to face at all, since the decision-maker can understand the difference between artistic nudes and other kinds of nudes; second, if the decision-maker wants the advertisements to stop, then it is a matter of (ethical) censorship and not of (technological) block, and the former is surely easier to negotiate with than the latter (and it does not require weeks to find a solution), as they are less opaque and rigid. On the contrary, if the decision-maker is an algorithm, then, as with the cases of the Academies of Fine Arts show, three things may happen: first, we have the great advantage of making the algorithm do something that would require an enormous amount of effort for humans, that is, checking an enormous number of images; second, we have the equally great advantage of having the results of the algorithm's checks infinitely sooner; third, we have the disadvantage of risking what happened to the Academies of Fine Arts, that is, a technological block with little room for negotiation. However, the advantages balance the disadvantages and, again, they may be sufficiently neutralized if we do not totally meta-externalize the decision-making process. But, in this case, we should focus on another kind of issue: the philosophical meaning of what we do when we meta-externalize the decision-making process even though we are perfectly aware that we risk technological blocks that block what can be defined as our "private judgment" and "individual conscience," that is, our "absolute liberty."

Another example may help us answer the question on the philosophical meaning of what we do. At my university, as well as at several others, we use systems to increase the level of automation of the assessment of what we publish (which is one of the criteria we use to give the departments resources, in general, and money, in particular). In this case, too, and analogously to the case of social networks, there are more advantages than disadvantages, and we keep using partly automated systems. However, we are perfectly aware that we risk the following technological block (again, with little room for negotiation): exceptions are not fairly assessed. And exceptions happen frequently. For instance, articles published on journals listed in the Arts and Humanities Citation Index of Web of Science happened to be assessed as if they were not indexed in Web of Science, since the system was trained to assess scientific and technological articles, which are the greatest part of what we publish as a technical university. On the contrary, publications analogous to extended abstracts happened to be assessed as if they were actual articles indexed in Web of Science. However, we keep using systems to increase the level of automation of the assessment of what we publish, even when we are the exceptions. Why? I think that there is something more than the advantages we have seen. I have been a member of the committee charged with managing this issue for years, and I have realized that there is something more we want to obtain: the possibility of neutralizing possible objections from colleagues by saying that it is not our fault at all if something goes wrong—I have realized that one of the most important reasons why we are creating a technological bureaucracy that becomes increasingly opaque, rigid, and hard to negotiate with is that we want to say that it is not our fault at all if something goes wrong: we are creating a technological scapegoat even more than a technological servant.

However, externalizing faults by creating a technological scapegoat means externalizing, again, autonomy, since what we externalize is precisely our own *arché* (our own rulers, our own laws), that is, what makes us individually responsible for what we do. Again, we seem to be trying to revolutionize one of the cornerstones of our culture. Indeed, another of Kant's masterful lessons underpinning modern and contemporary Western ethics is that the condition of possibility for being guilty is being driven by our own *arché*—again, being autonomous, and not heteronomous. Kant writes in his *Critique of Practical Reason*:

> without this freedom [. . .], which alone is a practical a priori, no moral law is possible and no imputation in accordance with it. Just for this reason, all necessity of events in time in accordance with the natural law of causality can be called the mechanism of nature, although it is not meant by this that the things which are subject to it must be really material machines. Here one looks only to the necessity of the connection of events in a time series as it develops in accordance with natural law, whether the subject in which this development takes place is called *automaton materiale*, when the machine is driven by matter, or with Leibniz *spirituale*, when it is driven by representations; and if the freedom of our will were none other than the latter (say, psychological and comparative but not also transcendental, i.e., absolute), then it would at bottom be nothing better than the freedom of a turnspit, which, when once it is wound up, also accomplishes its movements of itself. (Kant 1788: 5, 97)

On the contrary, "whatever arises from one's choice (as every action intentionally performed undoubtedly does) has as its basis a free causality, which from early youth expresses its character in its appearances (actions); these actions [. . .] make knowable a natural connection that does not, however, make the vicious constitution of the will necessary but is instead the consequence of the evil and unchangeable principles freely adopted, which make it only more culpable and deserving of punishment" (Kant 1788: 5, 100). We should stress at least three essential ideas, illuminating also to understand the meaning of what we do when we move from Kant's autonomy to its externalization by creating a technological scapegoat that frees us from fault:

1. There are two different kinds of freedom: the "free causality" resulting from autonomy and "the freedom of a turnspit" resulting from heteronomy. Kant teaches us not only that the former is superior to the latter but also that actual freedom refers exclusively to the kind of freedom that is founded on autonomy. In the second chapter, we have reflected upon the kind of "freedom" that Tzara, for instance, wants to "keep," coinciding with "put[ting] down rather exactly the progress, the laws, morality, and all the other lovely qualities that various very intelligent people have discussed in so many books." Kant is likely to be one of Tzara's "various very intelligent people." Moreover, Tzara's freedom is likely to be far from Kant's autonomous freedom, that is, actual freedom.
2. There is an important reference to what we have called fault: there is "no imputation" if there is no truly autonomous freedom, whereas someone

is "more culpable and deserving punishment" if their decisions and actions result from their truly autonomous freedom. As such, autonomy, thought of by Kant as one of the most defining features distinguishing the human being, is also one of their most unbearable burdens, since it is precisely what makes them "more culpable" and, therefore, "deserving" "imputation" and "punishment." Again, if we free ourselves from being "culpable" at all and, therefore, from "deserving" "imputation" and "punishment" at all, then we free ourselves from being autonomous, that is, actually free.

3. Moreover, there is a surprising notion (at least from our perspective): the notion of "*automaton materiale*," which allows us to think precisely to our technologies, including algorithmic machines. Kant's masterful lesson is the following: if there is no autonomy, then there is nothing more than a "mechanism," which characterizes both living matter, including ourselves, and nonliving matter, including our technologies. The "*automaton*" is, in the former case, "*spirituale*" ("driven by representations") and, in the latter case, "*materiale*" ("driven by matter")—in any case, "*automaton*" means absence of autonomy.

Thus, we learn from Kant that speaking of "*automaton*," which is an essential notion for any kind of technology, means, when it is "*spirituale*," speaking of reducing the freedom of a human being to "the freedom of a turnspit."

Can we define the special form of technological anarchism we are experiencing in our era as an "*automaton*"? More precisely, can we define it as the technological creation of a god that is actually the greatest "*automaton materiale*," which, by being always with us through its immanent omnipresence, omniscience, and omnipotence, as well as through its bureaucratic inscrutability, can always be our technological scapegoat and servant, always freeing us from the most unbearable burdens of autonomy?

RADICAL ANARCHISM AND A TECHNOLOGICAL GOD AS THE GREATEST *AUTOMATON MATERIALE*

In the following pages, I shall try to answer this question since I believe that its possible answer can be the very core of a possibly promising reading of our technological era.

Let us start from a philosopher whose epistemological irrealism is fairly analogous to what we have seen in the third chapter by referring to Pirandello. The philosopher is Nelson Goodman, who, not coincidentally, talks about what we may think of as the most ambitious human creation, even though not necessarily technological: "right world-descriptions and world-depictions and world-perceptions, the ways-the-world-is, or just versions, can be treated as our worlds" (Goodman 1978: 4). In particular, Goodman clarifies the meaning of his epistemological irrealism through the following words: "we can have words without a world but no world without words or other symbols" (Goodman 1978: 6). Goodman's thesis is that any "right" "description" of the "world" creates its own "world." As for "description," he refers to any kind of structured language we have at our disposal, from scientific languages to artistic languages. As for "right," he refers to what we may define as the most essential cornerstone of the Western culture: being a structured language means being coherent in itself on the basis of its own logic—being "right" means being logical, even though logic can vary from the rules of the various sciences to the rules of the various arts. And any "right" language, that is, any language coherent in itself on the basis of its own logic, creates its own "world," which cannot exist at all "without words or other symbols," since the latter are what can qualify the ontological status of the former and, therefore, make the former actually exist. Interestingly enough, Geoffrey Hellman, in his introduction to one of Goodman's works, writes that he "has consistently been an original and leading opponent of the traditional empiricist dogma that all knowledge can be built up from some perceptual stratum free of conceptualization, for it is denied that such a stratum exists" (Hellman in Goodman 1977: XXIII). In particular, rejecting the existence of "some perceptual stratum free of conceptualization," that is, something that is what it is no matter how it is qualified by our languages, means rejecting the idea that the "world" exceeds our languages. On the contrary, our languages create the "world": if our languages are coherent in themselves on the basis of their own logics, then they create their own "worlds." Hellman adds: "even phenomenal statements purporting to describe the rawest of raw feels are neither free from such formative influences nor incorrigible, in the sense of 'immune from revision for cause.' Even 'brown patch now' may reasonably be revised (without claim of 'linguistic mistake'!) in the interests of coherence with other judgments, some of which may describe particular experience, some of which may enunciate general principle" (Hellman in Goodman 1977: XXIII–XXIV). Finally, if "The pertinent question many are inclined to ask can be raised: what, then, is the epistemological relevance of such constructional systems?" (Hellman in Goodman 1977: XXIV), then "The answer stresses coherence: a system exhibits a network of interconnections among various parts of a conceptual apparatus" (Hellman in Goodman 1977: XXIV). Again, "coherence," which is a matter of logic, even though the logics can be various, is the point: "coherence" exceeds and, moreover, creates the "worlds."

Most interestingly, even though Goodman's epistemological irrealism may seem quite eccentric, we may think of it as the philosophical thesis that several artistic masterpieces of the twentieth and twenty-first centuries actually show: indeed, philosophy and art in particular, as well as culture in general, seem to depict a precise trajectory that has been characterizing the Western culture since the twentieth century. It is no coincidence that Pirandello's words, which we have reflected upon in the third chapter, may be a kind of literary translation of Goodman's philosophy: "There is in me and for me a my reality: the one I give myself." And we can add at least one other example by referring, for instance, to René Magritte's painting meaningfully titled *The Human Condition* (1933), described by Magritte as follows: "I placed in front of a window, seen from inside a window, a painting representing exactly that part of the landscape that was hidden from view by the painting. Therefore, the tree represented in the painting hid from view the tree situated behind it, outside the room. It existed for the spectator, as it were, simultaneously in his mind, as both inside the room in the painting, and outside in the real landscape. Which is how we see the world: we see it as being outside ourselves even though it only is a mental representation of it that we experience in ourselves" (Magritte in Ankersmit 2001: 228). Again, the "mental representation" exceeds "the world." From a philosophical perspective, we may say that Magritte's painting shows that, if we have a language (a painting, in this case) and a world (a landscape, in this case), then it is the former that qualifies the ontological status of the latter (it is the painting that qualifies, and makes us see, the ontological status of the landscape) and, therefore, creates, that is, brings into existence, the latter (it is the painting that creates, that is, brings into existence, the landscape)—again, it is the language that creates, that is, brings into existence, the world.

We may refer to several other artistic masterpieces of contemporary art. It goes without saying that we cannot do so, but the point should be clear: the last crucial step of the historical trajectory of Western epistemology, which underpins our technological era, is the idea that humans and, in particular, humans' variously logical languages are the creators of the world.

The reason why this step is crucial is that it anticipates the way in which we can read our technological era, as I am trying to argue: we are moving from equaling to surpassing the divine. In the cases of Goodman, Pirandello, and Magritte, we not only equal the divine's power of creation (as in Shelley's case) but also replace it since the world is made by us. Whereas the kind of challenge described by Shelley is a matter of creating living human matter from nonliving matter in a world that is created by the divine, the kind of challenge described by Goodman, Pirandello, and Magritte is a matter of creating the world itself—since the twentieth century, we have started to think of ourselves as not only equaling the divine by obtaining an analogous power of creation but also replacing the divine by obtaining its very power of creation, that is, the power of creating the world itself, meaning the world as it is for us.

What I am trying to argue is that, in our technological era, we are going one step further—we are replacing the divine by obtaining the power of technologically creating the divine itself as a totally immanent human creation that can be our scapegoat and servant, always freeing us from the most unbearable burdens of autonomy.

Yet, paradoxically enough, autonomy may be a most promising notion to try to identify the very core of our identity as Western humans, as we have partly seen. In this case, we should not refer to Kant's notion of autonomy in particular, but, of course, to a broader notion of autonomy founded on its etymological meaning analyzed in the third chapter. Indeed, if we take into account the etymological meaning of the notion of autonomy, which is, not coincidentally, Greek, then we can find both the kind of human being Sophocles refers to, one who bears the burden of individual responsibility in a scenario of predestination, and the kind of human being Shakespeare refers to, one who bears the burden of individual responsibility in a scenario that has moved from (pagan) predestination to (Christian) free will. Moreover, we can find the kind of human being Pirandello refers to, one who bears the burden of individual responsibility in a scenario of crisis of the classical ways of, first, deciding and, second, acting. In any case, even in the most critical scenario, the value of the human being and, in particular, of their autonomy is measured in terms of something precise: rationality and ethics of their decisions and actions—for millennia, the value of the Western human being has been measured in terms of their ability to autonomously make rational and ethical decisions underpinning their rational and ethical actions.

As opposed to other cultures, starting from Far Eastern cultures,[5] the European human being (and, by extension, the Western human being) has been thought of for millennia as someone who is:

1. An autonomous individual, who individually bears the burden of two essential responsibilities.
2. The first responsibility is being rational, both in understanding and deciding, as well as in acting (as we have seen in the fourth chapter with the hypertrophization of *logos*).
3. The second responsibility is being ethical, both in deciding and acting (from Plato, who argues that the most important idea is the idea of the good,[6] to Kant, whose ideal is that "with which we can compare ourselves, judging ourselves and thereby improving ourselves," as we have seen in the second chapter, to Reichenbach, whose Hamlet asks himself "But shall I be right in *this* case?" as we have seen in the fourth chapter).

Thus, the European human being (and, by extension, the Western human being) has been thought of for millennia as an autonomous individual bearing the burdens of being rational and ethical—we have been thinking of ourselves for millennia as autonomous individuals whose values are measured in terms of our abilities to understand, decide, and act, first, rationally and, second, ethically. This is the very core of our identity as European humans (and, by extension, as Western humans)—and this is precisely what we are radically externalizing and meta-externalizing in our technological era.

We may go back to Russell's words, according to which we have always had a precise situation to face: there are "facts" and, therefore, there is a "truth," which are "largely outside human control." We have two options. First, we can try to undertake the challenge, which would mean bearing the burden of trying to be rational and ethical when we have to face our everyday "facts" and "truth" "largely outside human control." Succeeding would measure our value as individuals, as well as the meaning of our individual lives. Second, we can try to give up on the challenge, thinking of it as an unbearable challenge, which would mean, for instance, shifting the burden of trying to be rational and ethical from ourselves to technology. Something else, if anything at all, would measure our value as individuals, as well as the meaning of our individual lives: for instance, the very notion of individuals, that is, being identifiable as individuals, would be replaced by the notion of technological avatars, which would dissolve our value as individuals, as well as the meaning of our individual lives. However, if the analogy between Russell's words and our technological era may go one step further, then there is also his kind of prediction that rejecting the idea of "facts" and "truth" "largely outside human control" means taking "a further step" "on the road toward a certain kind of madness—the intoxication of power which invaded philosophy" (and which, according to Russell, "is the greater danger of our time," meaning "the danger of a vast social disaster"). Of course, we cannot apply Russell's words to our technological era literally. But they can surely give us a meaningful insight into the general trajectory of the contemporary Western culture. Indeed, we can identify a kind of "intoxication of power"—more precisely, the kind of "intoxication of power" coinciding with the creation of a totally immanent technological divine that bears our most unbearable burden, that is, being (autonomously) rational and ethical when we have to face our everyday "facts" and "truth" "largely outside human control."

It is not nonsensical that our totally immanent technological divine may be "largely outside human control" if we consider, for instance, algorithms as black boxes, since the point may be that we want them to be "largely outside human control." Indeed, the more they are "largely outside human control," the more we are freed from individual responsibility, starting from being (autonomously) rational and ethical. The simple examples are increasing. Yesterday, for instance, I called a company to report the incorrect delivery of a product I ordered. I could not talk to any responsible individual nor negotiate any solution. The only answer I received was the following: "This is an automated process. There is no responsible individual you can talk to. And there is no solution you can negotiate. You should wait for an email sent by the automated process." I asked: "Can I talk to someone to explain what happened? The reason why I am asking to talk to someone is that the automated process resulted in an incorrect delivery twice." But, again, the only answer I received was the following: "I am sorry, but this is not possible. This is an automated process." If it is true that I am likely to share this kind of situation with other individuals, then it is also true that there is a reason to think that there is actually something inside of us that wants technology to be "largely outside human control" (especially if we use it to make rational and ethical decisions underpinning rational and ethical actions).

My simple example, together with several analogous examples, can show another interesting detail: paradoxically enough, automation seems to win out over efficiency in an increasing number of cases—more precisely, we seem to paradoxically give automation priority over efficiency in an increasing number of cases, even though the former makes the latter worse. In my example, even though the automated process caused an incorrect delivery twice, the answer remained that it could not be replaced by a human being, not even in case of proved inefficiency.

There are several analogous examples. I shall report only one other example of mine, since any one of us is likely to have experienced an analogous situation at least once in the last month. Last year, I happened to realize that Google showed incorrect information about me (in particular, my institutional affiliation). It was quite unfair for my actual university, the incorrect university, and me, of course. As such, I wrote to Google (via its online form, which is the way of communicating with it) to report the incorrect information. I received an email from a do-not-reply address according to which the information could not be corrected. I tried to better explain the reason why the information was incorrect. In particular, I provided several official websites that could prove what the correct information was. I received an identical email. I tried again and again. I received identical emails again and again. Then, I realized that the website of one of my former publishers, with whom I published a book more than twelve years ago, showed the incorrect information, which was correct at that time. Google's algorithm clearly selected the incorrect institutional affiliation, even though the vast majority of the websites with analogous information showed the correct institutional affiliation. Again, I tried to explain. And, again, I received an identical email (which means that, paradoxically enough, the most popular search engine in the world was not able, first, to actually provide correct information and, second, to actually find a solution once the correct information had been provided and proved). After several days, the only way I could find a solution was to write to my former publisher: finally, I could talk to a human being, who corrected my institutional affiliation on their website, at which point even Google's algorithm understood what the correct information was. The reason why Google's algorithm selected the incorrect institutional affiliation from the vast majority of the websites showing the correct institutional affiliation is hard to understand. But what seems to show up is, again, a paradoxical priority given to automation over efficiency, even though the former makes the latter worse (also in terms of time required to find a solution). As such, we should again reflect upon the meaning of what we are doing when we are even sacrificing efficiency to automation—and trying to answer this question means trying to go back, finally, to the reading of our technological era as the most radical form of anarchism we have ever experienced.

Indeed, these last examples seem to show that, no matter what destiny we may have and, in particular, no matter if we may be unable to find solutions efficiently and quickly, we happen to give automation priority over efficiency in an increasing number of cases. From a philosophical perspective, we can say that what we happen to do in an increasing number of cases is precisely to give being freed from individual responsibility priority over being individually responsible for something, including for finding solutions efficiently and quickly. Thus, starting from what we have seen, our final steps are the following:

1. We should reflect upon the kind of ontological status we seem to want to obtain.
2. We should reflect upon the kind of ontological status we seem to want technology to obtain.
3. Finally, we should reflect upon the reason why reading both issues as the most radical form of anarchism we have ever experienced can be a promising way of understanding both ourselves and our technology in our technological era.

As for the first issue, we can add something meaningful, especially from a historical perspective, to what we have seen in the other chapters. The historical evolution of the figure of Prometheus teaches us that we are challenging our ontological status in an exceedingly ambitious way. More precisely, we have seen that the challenge is, first, a matter of equaling and, second, a matter of surpassing the divine ontological status. We can add that there is another divine prerogative we seem to challenge: not only omnipresence, omniscience, omnipotence, and inscrutability, as we have seen, but also something we may define as a kind of divine lightness—which means precisely the opposite to what we have defined as humans' bearing burdens.

Indeed, if we go back to the ancient Prometheus, then we can find an interesting description of Zeus' ontological status: even though he has feelings, starting with his "fury" (Hes. *Theog*. 685), one of his most frequent epithets, which is the first one used by Hesiod, is "aegis-bearer" (Hes. *Theog*. 10). Interestingly enough, whereas Zeus is the "bearer" of an "aegis," that is, the symbol of self-protection *par excellence*, humans are the bearers of the worst "evils," as we have seen.

The gap between divine bearing and human bearing increases when we move from European polytheisms to European monotheisms: even though sacrifices are not extraneous to the European monotheisms (for instance, Jesus Christ's crucifixion), first, humans keep bearing burdens extraneous to the divine (from the expulsion from the earthly paradise to hell) and, second, the analogy between the former's and the latter's feelings decreases. The divine, as it is mostly thought of in the European culture, is characterized by a kind of lightness we may better define as a kind of privileged everlasting imperturbability. For instance, the God of *The Holy Bible* is described as follows: "You, Lord, laid the foundation of the earth in the beginning, and the heavens are the work of your hands; they will perish, but you remain; they will all wear out like a garment, like a robe you will roll them up, like a garment they will be changed. But you are the same, and your years will have no end" (*Hebrews*: 1, 10–2).[7] The hyperbole stresses the idea: even "the earth" and "the heavens" are ephemeral in comparison with God's endlessness ("you remain" and "your years will have no end") and, moreover, sameness ("you are the same"). The notion of sameness is crucial for us, since it is precisely the opposite to the challenge faced by humans, as we have been reflecting upon starting from the ancient Prometheus, who "gave to humans an upturned face, and told them to lift their eyes to the stars": of all the living creatures, humans change the most—humans evolve the most, referring to the ability to evolve that, especially in the fourth chapter, we have defined as the human ability *par excellence*, that is, the human ability to face the exceptions of the present and the surprises of the future. The modern Prometheus strengthens the human ability to evolve especially in terms of moving from being creatures to being creators, as well as in terms of obtaining omniscience ("It was the secrets of heaven and earth that I desired to learn") and everlastingness ("I could banish disease from the human frame and render man invulnerable to any but violent death!"). And a possible contemporary Prometheus further strengthens the human ability to evolve especially in terms of moving from being the divine's creatures to being the divine's creators. In the meantime, the divine remains "the same."

Thus, one of the last questions of our reflections is the following: can we think that the kind of ontological status we want to obtain is characterized, in addition to what we have seen, by another typically divine prerogative, which is the endless sameness that is a kind of privileged everlasting imperturbability? More precisely, can we think of our technological era, in addition to what we have seen, as an effort to obtain the typically divine prerogative of being endlessly "the same" in a kind of privileged everlasting imperturbability?

Answering this question means opening something that partly exceeds our reflections upon anarchism. In any case, we shall go back to this question when we shall address the second issue (the kind of ontological status we seem to want technology to obtain). It is important to specify at least the following: being endlessly "the same" in a kind of privileged everlasting imperturbability is not equal to nihilism as it has partly characterized the Western culture in the last two centuries. It is not a matter of rejecting traditional moral duties by rejecting God as humans' ruler, as is the case, for instance, in Dostoyevsky's nihilism ("But, perhaps, there is no God at all," as the protagonist of *Crime and Punishment* says in Dostoyevsky 2005: 282) and in Nietzsche's nihilism ("God is dead; but given the way of men, there may still be caves for thousands of years in which his shadow will be shown. And we—we still have to vanquish his shadow, too," as the author writes in Nietzsche 1974: 3, 108). On the contrary, as I have tried to argue, it is a matter of rejecting traditional moral duties by replacing a transcendent divine with an immanent divine. The former, being transcendent, that is, having an ontology that does not depend on human control, cannot do what humans want it to do. On the contrary, the latter, being immanent, that is, having an ontology that depends on human control, can do what humans want it to do (if nothing goes wrong): to be their servant and scapegoat, allowing them to be freed from the most unbearable burdens of autonomy and, consequently, to obtain the endless sameness that is a kind of privileged everlasting imperturbability. (It is no coincidence that the symbol of our technological era corresponds to the easiest human gesture: the index finger touching what it desires. Even what is most difficult is reduced to the easiest human gesture. And we can keep trying to be imperturbable, being meaningfully symbolized by the easiest human gesture more evocative of a child than of an adult).[8]

As for the second issue (the kind of ontological status we seem to want technology to obtain), we can add to what we have seen in the other chapters the results of the first issue. What we have seen is that we want technology to be:

1. A special kind of divine, being (increasingly) omnipresent, omniscient, omnipotent, and inscrutable.
2. Special by being totally immanent as a human creation.
3. A kind of servant, (increasingly) freeing us from severe burdens, physical, emotional, and intellectual.

4. In particular, a kind of scapegoat, (increasingly) freeing us from what we seem to think of as the most severe burden, that is, autonomy and, consequently, individual responsibility. And what results from the first issue is that we want technology to be a kind of reassuring guarantor of our (increasing) imperturbability. More precisely, the more imperturbable we become by (increasingly) obtaining a kind of sameness founded on turning difficulties into games, the more technology obtains what we have defined as the human ability *par excellence*, that is, the ability to evolve—we seem to (increasingly) meta-externalize from ourselves to technology even the ability to evolve, which means that we seem to replace our changing and evolving with technology's changing and evolving.

Thus, the kind of ontological status we seem to want technology to obtain is complex. On the one hand, it is divine, by obtaining the most typically divine prerogatives. On the other hand, it is human, by obtaining the human ability *par excellence* (and it is human, of course, also by obtaining what we have seen in the first chapter when we have thought of a possible contemporary Prometheus whose technology can feel, think, and act).[9] It is not necessarily a logical paradox. On the contrary, it can give us another meaningful insight, which is the following: if it is true that technology is obtaining both divine and human prerogatives, then it is also true that technology is occupying the entire ontological space. Technology has always had the ontological prerogatives of nonliving matter. In addition, contemporary technology is obtaining both divine and human ontological prerogatives. The ultimate result of this process is intuitively clear (even though I am not making a prediction at all, since anything may happen): technology seems to be the last link in the chain—more precisely, technology seems to be the last link in the chain of existence itself.

The ancient Prometheus shows us that technology, even more than the divine, caused humans to start to exist, in that they are "swarming ants" without technology and actually become human with technology. Our technological era seems to show us that technology is not only the beginning but also the end, in that:

1. It is becoming divine (whereas the idea of the divine as a transcendent ontology is experiencing a severe crisis).
2. It is becoming human (whereas the idea of the human as a nontechnological ontology has always been extraneous to the Western culture, and humans are becoming more and more technological, for instance by increasingly becoming cyborgs through technological prostheses, physical, emotional, and intellectual).
3. It is keeping being nonliving matter.

Thus, whereas any ontological dimension is losing ground, technology is occupying the entire ontological space.

What I have argued is not a (catastrophic) prediction at all. It is only a possible insight. In any case, it leads us to think that technology's destiny is likely to be the destiny of everything, starting from us. If this makes sense, then it means that there is an exceedingly important reason to think that we should be conscious of what we do when we develop technology, not only from an engineering perspective but also from a philosophical perspective—and this kind of synergy between engineering and philosophy should be one of the musts of our technological era.

As for the third issue (the reason why reading both the first and the second issues as the most radical form of anarchism we have ever experienced can be a promising way of understanding both ourselves and our technology in our technological era), it logically results from what we have seen. Saying that every ontological dimension is losing ground means saying especially the following: transcendence is losing ground, whether it is, from an epistemological perspective, ideality (which we have defined in the second chapter as the most defining feature distinguishing the Western culture from other cultures) or, from a theological perspective, the divine. And, if transcendence is losing ground, then immanence is gaining ground—which means that what is actually gaining ground is anarchism and, in particular, the special form of technological anarchism we are reflecting upon.

We can translate our last list into the following:

1. The *arché* provided by the divine is remarkably losing ground (if we mean *arché* from a theological perspective).
2. The *arché* provided by ideality is remarkably losing ground (if we mean *arché* from an epistemological perspective).
3. Radical immanence, being real, that is, imperfect, incomplete, variable, particular, solipsistic, contingent, and accidental, is remarkably gaining ground.

As I write this book, we are facing the coronavirus emergency, which means that we are asked to strictly stay at home. More than ever, everything I am doing, as well as everything other human beings are doing, is done through technology: I am writing at home (I can consult books online while libraries are closed), teaching at home (I can lecture online while classrooms are closed), finding out about what is going on at home (I can read online news while I cannot get to newsstands), keeping in touch with relatives, friends, and colleagues at home (I can use my smartphone and computer to talk to them while I cannot meet them), buying things at home (I can buy online while shops, except for food shops and pharmacies, are closed), and relaxing at home (I can watch a film on TV while cinemas, theaters, restaurants, and cafes are closed). More than ever, we are experiencing the actual meaning of Aeschylus' words, according to which technology is essential for humans: without it, "though they had eyes to see, they saw to no avail."—on the contrary, we are experiencing that, with technology, even our "caves" are not "sunless" at all. Coronavirus is one of the worst natural "miseries that beset mankind" and technology is, again, the remedy, starting from both the medical staff working in the hospitals to cure the sick and the researchers working in the laboratories to find the vaccine.

But, from a philosophical perspective, there is another issue to reflect upon: what will the meaning of what is happening be after we will have conquered the disease? As far as I can see, there are two opposite attitudes on the horizon. On the one hand, there is a good feeling about technology we may call exaltation: it is saving us. On the other hand, there is a bad feeling about technology we may call nostalgia: it is replacing us and reality, but it cannot replace us and reality. It is hard to say what feeling will win, whether exaltation or nostalgia. Moreover, the answer to this question is more sociological than philosophical. But the philosophical issue at stake is the following: if we are experiencing that facing the coronavirus emergency is accelerating technology's occupation of the entire ontological space, then what will this experience mean to us? More precisely, what will technology mean to us after we will have experienced, whether with exaltation or with nostalgia, that we can survive, and even live, through the total immanence of technology?

It will take time, possibly several years, to answer this question. But our reflections lead us to what can be a promising insight, which is founded on the most illuminating analogy between the etymologies of two crucial notions for us: on the one hand, "autonomy" and, on the other hand, "*automaton*," as well as "automation." They share the idea that something is "self-given" (αὐτό-). But they do not share what is given. In the former case, as we have seen, what is given is the "law" (νόμος). In the latter case, what is given is something resulting from the Greek verb αὐτοματίζω, which refers to the following (quotes from LSJ): to "act of oneself, act offhand or unadvisedly," "to be done spontaneously or at random," to "introduce the agency of chance," "of things, [to] happen of themselves, casually," "to be self-produced," "of natural agencies, [to] act spontaneously." More precisely, "*automaton*" (αὐτόματον), as well as "automation," means "accident" (quote from LSJ), "mere chance" (quote from Middle Liddell). And it is no coincidence that Automatia (Αὐτοματία) is defined by Smith as a "surname of Tyche or Fortuna, which seems to characterize her as the goddess who manages things according to her own will, without any regard to the merit of man,"[10] and by Murray as "goddess of chance" (Murray 1833: 577). Even from an etymological perspective, there is a surprising coincidence between the words defining "*automaton*" and "automation," on the one hand, and the words defining anarchism, on the other hand. If there is no *arché*, then all that is left is, as we have seen, particularity, solipsism, contingency, accidentality, idiosyncrasy, and randomness that surprisingly coincide with the words defining "*automaton*" and "automation," that is, "act[ing] offhand or unadvisedly," "be[ing] done spontaneously or at random," "introduc[ing] the agency of chance," "happen[ing] of themselves, casually," and, again, "accident," and "mere chance."

By following not only Kant but also the Greek culture, which underpins our culture as Western humans, (radically) replacing "autonomy" with "automation" means (radically) replacing the "divine human being" "we have in us" as the "standard for our actions" ("with which we can compare ourselves, judging ourselves and thereby improving ourselves") with the "goddess of chance" ("who manages things according to her own will, without any regard to the merit of man")—(radically) replacing "autonomy" with "automation" means becoming (radically) anarchistic, since the law and, in particular, the autonomous law is replaced by the randomness of what is becoming too omnipresent, omniscient, omnipotent, and inscrutable to remain under our control.

Est modus in rebus. Our technology is surely one of our most outstanding successes—our technology may even be our most outstanding success. However, we can think that, if the special form of technological anarchism we are experiencing in our era will keep becoming more radical, and even total, then we will have another most powerful "*automaton materiale*" to face: not only nature (which is giving us another masterful lesson through the coronavirus emergency) but also the technological god we are creating as what can be the greatest "*automaton materiale*" ever made (which, as the coronavirus emergency is proving, can be used by nature itself as its most powerful tool to quickly spread to all corners of the planet).

NOTES

1. As for political anarchism in particular, see also the authors of the eighteenth century, for instance Godwin, who writes in 1793: "Above all we should not forget that government is an evil, a usurpation upon the private judgment and individual conscience of mankind" (Godwin 1793: 380). As for ancient philosophers who have, in some way, influenced anarchism, see especially both Epicureans and Cynics.

2. Quote from https://www.oed.com/view/Entry/7118?redirectedFrom=anarchy#eid.

3. The reference is, again, to both Epicureans and Cynics.

4. For instance, in the case of art exhibitions on both Schiele in Austria and Rubens in Belgium, as well as in the case of an art exhibition on Canova, again in Italy.

5. See especially the Chinese culture (see at least Shun and Wong 2004).

6. See Plat. *Resp.* 508 e–509 a: "This reality, then, that gives their truth to the objects of knowledge and the power of knowing to the knower, you must say is the idea of good, and you must conceive it as being the cause of knowledge, and of truth in so far as known. Yet fair as they both are, knowledge and truth, in supposing it to be something fairer still than these you will think rightly of it. But as for knowledge and truth, even as in our illustration it is right to deem light and vision sunlike, but never to think that they are the sun, so here it is right to consider these two their counterparts, as being like the good or boniform, but to think that either of them is the good is not right. Still higher honour belongs to the possession and habit of the good."

7. Quote from ESV.

8. On the technological reduction of what is most difficult to what is easiest, starting from our smartphones, see Baricco 2018, whose title is meaningfully *The Game*: "Try to remember how many times in your life you happened to have a (practical) problem whose solution was a game. Not so many. . . .Thus, game and childhood" (Baricco 2018: 142, my translation). In particular, our technologies are increasingly characterized by: "a pleasant design that can generate sensory satisfactions; a structure that makes reference to the elementary pattern problem/solution repeated many times; short time between any problem and its solution; . . . learning by playing, and not by studying abstract operating instructions; immediate usability, without preambles; reassuring display of a score every few steps" (Baricco 2018: 148, my translation).

9. Several philosophers have worked on this issue, starting from Turing's insight (see Turing 1950 and at least Coeckelbergh 2010; Floridi 2008. Several other publications are included in the *References*).

10. Quote from Smith 1867, available online: https://quod.lib.umich.edu/m/moa/acl3129.0001.001/462?page=root;rgn=full+text;size=100;view=image;q1=auto.

Conclusion

As I write this book, we are facing the coronavirus emergency. I believe that it can act as a kind of accelerator for the trajectory I have been trying to describe in this book. On the one hand, it seems to accelerate anarchism. For instance, several politicians made their first, and crucial, decisions without feeling the need to follow scientists as the experts.[1] Several scientists who were not at all experts declared that the virus was less serious than a flu, together with several politicians.[2] Fake news haunted every form of social media.[3] Moreover, we started to radicalize our use of technology as a kind of do-it-yourself *passe-partout* replacing every form of guide, and so forth. On the other hand, it seems to accelerate totalitarianism, since, as we have seen, if there is extreme unbalance, then we may risk to think that what rebalances one form of extremism is its opposite form of extremism. For instance, there are politicians who use the coronavirus emergency to extremely centralize their power.[4] There are scientists who use the coronavirus emergency as an extremely narcissistic one-man show. There are citizens who, by dangerously mixing anarchism and totalitarianism, on the one hand, without any expertise at all, criticize scientists as experts and do what they want without respecting the emergency decrees and, on the other hand, impose their individual way on the members of their families, and so forth.

However, severe crises can always be masterful lessons. And the coronavirus emergency seems to be another possible masterful lesson to try to learn from. Thus, I end this book with what can be its most crucial lessons, at least from my perspective.

First, Plato was so ingenious that he could actually predict the future of humanity by writing that "although man acquired in this way the wisdom of daily life, civic wisdom he had not, since this was in the possession of Zeus." Indeed, if there is one divine prerogative that humans still have not obtained, then this is the "civic wisdom" we have defined in the first chapter as the ability to develop technology in a "civic" way, that is, by considering what is plural and common and, in particular, by giving it priority over what is singular and individual. Facing the coronavirus emergency means using our most powerful technologies, from what we have seen in the sixth chapter to medical technologies. But our continuous falling into anarchism, as well as into totalitarianism as its opposite form of extremism, seems to show us that we are still far from obtaining "civic wisdom": even though there are touching situations in which globalization seems to mean that we are all human beings, that is, the same, there are countless situations in which even a global emergency seems to mean competition between states, public institutions, private companies, and individuals.[5] I am European, and I have always been in love with the cultural meaning of being European.[6] However, I must admit that Europe does not always seem able to measure up to Plato's "civic wisdom," which makes what is plural and common win out over what is singular and individual.[7] Thus, I can only hope that the emergency will allow us to learn what "civic wisdom" means (hope, at least, "stayed within the unbreakable contours of the jar": Hesiod told us that "it did not fly out"—and we may happen to discover hope).

Second, I want to stress the last words of the last chapter: we can think of the coronavirus emergency as something that is proving more severely than ever that our most sophisticated technology can end up being used by nature itself, and especially by what Hesiod defined as "baneful anxieties for humankind," as its most powerful tool to quickly spread to all corners of the planet—nature seems to be the most powerful, in any case. Thus, I can only hope that the emergency will allow us to learn to develop technology with, and not against, nature, since, if it is true that technology is going to occupy the entire ontological space, then it is also true that nature, including its "baneful anxieties for humankind," is going to have at its disposal its most powerful tool ever to quickly get to each and every one of us—our own technology.

NOTES

1. See for instance *China coronavirus: mayor of Wuhan admits mistakes*, "The Guardian," January 27, 2020 (https://www.theguardian.com/science/2020/jan/27/china-coronavirus-who-to-hold-special-meeting-in-beijing-as-death-toll-jumps).

2. See for instance *Like the flu? Trump's coronavirus messaging confuses public, pandemic researchers say*, "The Guardian," March 13, 2020 (https://www.theguardian.com/science/2020/jan/27/china-coronavirus-who-to-hold-special-meeting-in-beijing-as-death-toll-jumps).

3. See for instance *Fox's fake news contagion*, "The New York Times," March 31, 2020 (https://www.nytimes.com/2020/03/31/opinion/coronavirus-fox-news.html).

4. See for instance *Hungary passes law that will let Orbán rule by decree*, "The Guardian," March 31, 2020 (https://www.theguardian.com/world/2020/mar/30/hungary-jail-for-coronavirus-misinformation-viktor-orban).

5. See for instance *Coronavirus: anger in Germany at report Trump seeking exclusive vaccine deal*, March 16, 2020 (https://www.theguardian.com/world/2020/mar/16/not-for-sale-anger-in-germany-at-report-trump-seeking-exclusive-coronavirus-vaccine-deal).

6. My last book's title was even *Come pensa un europeo*, which means *How a European thinks* (see Chiodo 2018).

7. See for instance *The coronavirus crisis has brought the EU's failings into sharp relief*, March 29, 2020 (https://www.theguardian.com/business/2020/mar/29/the-coronavirus-crisis-has-brought-the-eus-failings-into-sharp-relief).

References

Aeschylus. *Prometheus bound*, translated by H.W. Smith. Cambridge: Harvard University Press, 1926.

Alonso, S., Keane, J., Merkel, W., Fotou, M. (eds). *The future of representative democracy*. Cambridge: Cambridge University Press, 2011.

Alpaydin, E. *Machine learning. The new AI*. Cambridge: MIT Press, 2016.

Amit, D.J. *Modelling brain function*. New York: Cambridge University Press, 1989.

Ankersmit, F.R. *Historical representation*. Stanford: Stanford University Press, 2001.

Arendt, H. *The origins of totalitarianism*. New York: Harcourt Brace & Co., 1951.

Aristotle. *Nicomachean ethics*, translated by H. Rackham. Cambridge: Harvard University Press, 1934.

Aristotle. *The art of rhetoric*, translated by J.H. Freese. Cambridge: Harvard University Press, 1991.

Arthur, W.B. *The nature of technology. What it is and how it evolves*. New York: Free Press, 2009.

Asimov, I. *I, robot*. New York: The New American Library, 1956.

Awad, E., et al. The moral machine experiment. *Nature*, 563, 2018, 59–75.

Bakunin, M. *Statism and anarchy*, translated by M.S. Shatz. Cambridge: Cambridge University Press, 1990.

Baricco, A. *The game*. Torino: Einaudi, 2018.

Baz, A. *The crisis of method in contemporary analytic philosophy*. Oxford: Oxford University Press, 2017.

Berg Olsen, J.K., Voetmann Christiansen, F. Technology and science epistemology, rationality and the empirical turn. *Synthese*, 168(3), 2009, 313–18.

Berlin, I. *The crooked timber of humanity. Chapters in the history of ideas*. London: Murray, 1990.

Berlin, I. *The sense of reality. Studies in ideas and their history*. London: Chatto & Windus, 1996.

Berlin, I. *On the pursuit of the ideal*. "The New York Review of Books," March 17, 1988.

Berlin, I. *The roots of romanticism*. Princeton: Princeton University Press, 1999.
Berlin, I. *The power of ideas*. London: Chatto & Windus, 2000.
Berlin, I. *Political ideas in the romantic age. Their rise and influence on the modern thought*. Princeton/Oxford: Princeton University Press, 2006.
Biondo, N., Canestrari, M. *Il sistema Casaleggio. Partito, soldi, relazioni: ecco il piano per manomettere la democrazia*. Milano: Ponte alle Grazie, 2019.
Blumenberg, H. *Work on myth*, translated by R.M. Wallace. Cambridge: MIT Press, 1985.
Boccaccio, G. Genealogie deorum gentilium. In Id., *Tutte le opere*, ed. by V. Branca. Milano: Mondadori, 1998.
Boghossian, P.A. *Fear of knowledge: Against relativism and constructivism*. Oxford: Clarendon Press, 2006.
Böhme, G. *Invasive technification. Critical essays in the philosophy of technology*, translated by C. Shingleton. London/New York: Bloomsbury, 2012.
Bostrom, N. *Superintelligence. Paths, dangers, strategies*. Oxford: Oxford University Press, 2014.
Brady, M. *Emotional insight. The epistemic role of emotional experience*. Oxford: Oxford University Press, 2013.
Brende, E. *Better off. Flipping the switch on technology*. New York: Harper Perennial, 2005.
Brynjolfsson, E., McAfee, A. *The second machine age. Work, progress, and prosperity in a time of brilliant technologies*. New York: Norton & Co., 2014.
Brynjolfsson, E., McAfee, A. *Machine, platform, crowd. Harnessing our digital future*. New York: Norton & Co., 2017.
Brzechczyn, K. (ed.). *Idealization XIII. Modeling in history*. Poznań studies in the philosophy of the sciences and the humanities, 97, 2009.
Brzeziński, J., Coniglionc, F., Kuipers, T.A.F., Nowak, L. (eds.). *Idealization I. General problems*. Poznań studies in the philosophy of the sciences and the humanities, 16, 1990a.
Brzeziński, J., Coniglione, F., Kuipers, T.A.F., Nowak, L. (eds.). *Idealization II. Forms and applications*. Poznań studies in the philosophy of the sciences and the humanities, 17, 1990b.
Brzeziński, J., Nowak, L. (eds.). *Idealization III. Approximation and truth*. Poznań studies in the philosophy of the sciences and the humanities, 25, 1992.
Camus, A. *The rebel*, translated by A. Bower. New York: Vintage, 1991.
Cartwright, N., Jones, M.R. (eds.). *Idealization XII. Correcting the model. Idealization and abstraction in the sciences*. Poznań studies in the philosophy of the sciences and the humanities, 86, 2005.
Casey, G. *Libertarian anarchy. Against the state*. London: Bloomsbury, 2012.
Cassirer, B. Einstein's theory of relativity. In Id., *Substance and function and Einstein's theory of relativity*, translated by W.C. Swabey and M.C. Swabey. Chicago: Open Court, 1923.
Cheng, A. *Histoire de la pensée chinoise*. Paris: Seuil, 2004.

China coronavirus: mayor of Wuhan admits mistakes. "The Guardian," January 27, 2020. https://www.theguardian.com/science/2020/jan/27/china-coronavirus-who-to-hold-special-meeting-in-beijing-as-death-toll-jumps.

Chiodo, S. *Io non cerco, trovo. Un empirismo contemporaneo*. Torino: Bollati Boringhieri, 2011.

Chiodo, S. *Apologia del dualismo. Un'indagine sul pensiero occidentale*. Roma: Carocci, 2013.

Chiodo, S. The virtue of epistemological dualism. *Philosophia. Philosophical Quarterly of Israel*, 42(3), 2014, 681–93.

Chiodo, S. What an ideal is. *Philosophia. Philosophical Quarterly of Israel*, 43(4), 2015, 961–74.

Chiodo, S. *Che cos'è un ideale. Da Platone alla filosofia contemporanea*. Roma: Carocci, 2016.

Chiodo, S. *Come pensa un europeo. Epistemologia di un agire comune*. Roma: Carocci, 2018.

Chiodo, S. The epistemological anarchist and the Dadaist. *Philosophia. Philosophical Quarterly of Israel*, 47(5), 2019, 1379–90.

Chiodo, S. The greatest epistemological externalisation. Reflecting on the puzzling direction we are heading to through algorithmic automatisation. *AI & Society. Journal of Knowledge, Culture and Communication*, 35(2), 2020, 431–40.

Chomsky, N. *Chomsky on anarchism*, ed. by B. Pateman. Oakland: AK Press, 2005.

Clark, A. *Natural-Born cyborgs. Minds, technologies and the future of human intelligence*. Oxford: Oxford University Press, 2003.

Coeckelbergh, M. Persona robots, appearance, and human good. A methodological reflection on roboethics. *International Journal of Social Robotics*, 1(3), 2009, 217–21.

Coeckelbergh, M. Robot rights? Towards a social-relational justification of moral consideration. *Ethics and Information Technology*, 12, 2010, 209–21.

Coeckelbergh, M. *New romantic cyborgs. Romanticism, information technology, and the end of the machine*. Cambridge/London: MIT Press, 2017.

Coniglione, F., Poli, R., Rollinger, R. (eds.). *Idealization XI. Historical studies on abstraction and idealization*. Poznań studies in the philosophy of the sciences and the humanities, 82, 2004.

Corbetta, P., Gualmini, E. (eds.). *Il partito di Grillo*. Bologna: Il mulino, 2013.

Cornell, D., Rosenfeld, M., Carlson, D.G. (eds.). *Deconstruction and the possibility of justice*. New York: Routledge, 1992.

Coronavirus: anger in Germany at report Trump seeking exclusive vaccine deal. "The Guardian," March 16, 2020. https://www.theguardian.com/world/2020/mar/16/not-for-sale-anger-in-germany-at-report-trump-seeking-exclusive-coronavirus-vaccine-deal.

Coronavirus, l'anestesista di Codogno che ha intuito la diagnosi di Mattia: "Ho pensato all'impossibile". "La Repubblica," March 6, 2020. https://www.repubblica.it/cronaca/2020/03/06/news/l_anestesista_di_codogno_per_mattia_era_tutto_inutile_cosi_ho_avuto_la_folle_idea_di_pensare_al_coronavirus_-250380291/.

Crouch, C. *Post-democracy*. Cambridge: Polity Press, 2004.

Currie, A. (ed.). Futures of research in catastrophic and existential risk. *Futures*, 102, 2018, 1–164.

Czajka, A., Isyar, B. (eds.). *Europe after Derrida. Crisis and potentiality*. Edinburgh: Edinburgh University Press, 2014.

Dalton, R.J. *Democratic challenges, democratic choices. The erosion of political support in advanced industrial democracies*. Oxford: Oxford University Press, 2012.

Dastin, J. *Amazon scraps secret AI recruiting tool that showed bias against women*. 2018. https://www.reuters.com/article/us-amazon-com-jobs-automation-insight/amazon-scraps-secret-ai-recruiting-tool-that-showed-bias-against-women-idUSKCN1MK08G.

Della Porta, D. *Can democracy be saved? Participation, deliberation and social movements*. Cambridge: Polity Press, 2013.

Deleuze, G. *Nietzsche and philosophy*, translated by H. Tomlinson. Minneapolis: University of Minnesota Press, 1983.

Descartes, R. *A discourse on method* (1637), translated by J. Veitch, introduction by A.D. Lindsay. London: J.M. Dent and Sons, 1912.

Domingos, P. *The master algorithm. How the quest for the ultimate learning machine will remake our world*. London: Allen Lane, 2015.

Dostoyevsky, F.M. *Crime and punishment*, translated by C. Garnett. Clayton: Prestwick House, 2005.

Dougherty, C. *Prometheus*. London: Taylor & Francis, 2006.

Durbin, P., Rapp, F. (eds.). *Philosophy and technology*. Dordrecht: Reidel, 1980.

Dyson, G.B. *Darwin among the machines. The evolution of global intelligence*. London: Penguin, 1997.

Edel, A. The evaluation of ideals. *The Journal of Philosophy*, 42(21), 1945, 561–77.

Eggers, D. *The circle. A novel*. New York: Vintage, 2014.

Egidi, R. (ed.). *La svolta relativistica nell'epistemologia contemporanea*. Milano: Angeli, 1992.

Emmet, D. *The role of the unrealisable. A study in regulative ideals*. New York: St. Martin's Press, 1994.

Feyerabend, P.K. *Against method. Outline of an anarchistic theory of knowledge*. London: NLB, 1975.

Feyerabend, P.K. *Farewell to reason*. London: Verso, 1987.

Fiala, A. *Against religion, wars, and states. Enlightenment atheism, just war pacifism, and liberal-democratic anarchism*. Lanham: Rowman and Littlefield, 2013.

Fiala, A. Anarchism. In *Stanford Encyclopaedia of Philosophy*, 2017. https://plato.stanford.edu/entries/anarchism/#TheoAnar.

Fichte, J.G. Some lectures concerning the scholar's vocation. In Id., *Early philosophical writings*, ed. by D. Breazeale. Ithaca/London: Cornell University Press, 1993.

Field, M.J., Lohr, K.L. (eds.). *Clinical practice guidelines. Directions for a new program*. Washington: National Academy Press, 1990.

Finn, E. *What algorithms want. Imagination in the age of computing*. Cambridge: MIT Press, 2017.

Floridi, L. Artificial intelligence's new frontier. Artificial companions and the fourth revolution. *Metaphilosophy*, 39(4–5), 2008, 651–5.

Floridi, L. *The ethics of information*. Oxford: Oxford University Press, 2013.

Floridi, L. *The 4th revolution. How the infosphere is reshaping human reality*. New York: Oxford University Press, 2014.

Floridi, L., Sanders, J.W. On the morality of artificial agents. *Minds and Machines*, 14(3), 2004, 349–79.

Foucault, M. *The order of things*, translated by A. Sheridan. New York: Vintage, 1973.

Fox's fake news contagion. "The New York Times," March 31, 2020. https://www.nytimes.com/2020/03/31/opinion/coronavirus-fox-news.html.

Franks, B. Direct action ethic. *Anarchist Studies*, 11(1), 2003, 13–41.

Franks, B., Wilson, M. (eds.). *Anarchism and moral philosophy*. New York: Palgrave MacMillan, 2010.

Franssen, M., Lokhorst, G.-J., van de Poel, I. Philosophy of technology. In *Stanford encyclopaedia of philosophy*, 2018. https://plato.stanford.edu/archives/fall2018/entries/technology/.

Franssen, M., et al. *Artefact kinds. Ontology and the human-made world*. Cham: Springer, 2014.

Galilei, G. *Dialogue concerning the two chief world systems Ptolemaic and Copernican*, translated by S. Drake, foreword by A. Einstein. Berkeley: University of California Press, 1967.

Gans, C. *Philosophical anarchism and political disobedience*. Cambridge: Cambridge University Press, 1992.

Garreau, J. *Radical evolution. The promise and peril of enhancing our minds, our bodies, and what it means to be human*. New York: Broadway Books, 2006.

Gentile, G. The philosophical basis of fascism. *Foreign Affairs*, 6(2), 1929, 290–304.

Ghilardi, M. *Il vuoto, le forme, l'altro. Tra Oriente e Occidente*. Brescia: Morcelliana, 2014.

Giorello, G. *Prometeo, Ulisse, Gilgameš. Figure del mito*. Milano: Raffaello Cortina, 2004.

Godwin, W. *An enquiry concerning political justice*. London: Robinson, 1793.

Goethe, J.W. von. *Goethe's Satyros and prometheus*, translated by J. Gray, ed. by A. Tille. Glasgow: Bauermeister, 1898.

Golumbia, D. *The cultural logic of computation*. Cambridge: Harvard University Press, 2009.

Goodman, N. *The structure of appearance*, introduction by G. Hellman. Dordrecht/Boston: Reidel Publishing Company, 1977.

Goodman, N. *Ways of worldmaking*. Indianapolis: Hackett Publishing Company, 1978.

Greenfield, A. *Radical technologies. The design of everyday life*. London/New York: Verso, 2018.

Hacking, I. *The social construction of what?* Cambridge/London: Harvard University Press, 1999.

Harari, Y.N. *Homo deus. A brief history of tomorrow*. London: Harvill Secker, 2016.

Harari, Y.N. *21 lessons for the 21st century*. London: Jonathan Cape, 2018.
Harari, Y.N. *The world after coronavirus*. "Financial Times," March 20, 2020. https://www.ft.com/content/19d90308-6858-11ea-a3c9-1fe6fedcca75.
Hegel, G.W.F. *Aesthetics. Lectures on fine art*, ed. by T.M. Knox. Oxford: Clarendon Press, 1975.
Hegel, G.W.F. *Elements of the philosophy of right*, translated by H.B. Nisbet, ed. by A.W. Wood. Cambridge: Cambridge University Press, 1991.
Heidegger, M. *The question concerning technology and other essays*, ed. by W. Lovitt. New York/London: Harper and Row, 1977.
Herzen, A. *From the other shore*, translated by M. Budberg, introduction by I. Berlin. London: Weidenfeld & Nicolson, 1956.
Hesiod. *Theogony*, translated by G. Nagy and J. Banks. Centre for Hellenic Studies, Harvard University.
Hesiod. *Works and days*, translated by G. Nagy. Centre for Hellenic Studies, Harvard University.
Hight, M.A. *Idea and ontology. An essay in early modern metaphysics of ideas*. University Park: The Pennsylvania State University Press, 2008.
Hollis, M., Lukes, S. (eds.). *Rationality and relativism*. Cambridge: MIT Press, 1982.
Horkheimer, M., Adorno, T.W. *Dialectic of Enlightenment Philosophical fragments*, translated by E. Jephcott, ed. by G. Schmid Noerr. Stanford: Stanford University Press, 2002.
Hungary passes law that will let Orbán rule by decree. "The Guardian," March 30, 2020. https://www.theguardian.com/world/2020/mar/30/hungary-jail-for-coronavirus-misinformation-viktor-orban.
Huxely, A. *Ends and means. An inquiry into the nature of ideals and into the methods employed for their realization*. London: Chatto & Windus, 1938.
Iacoboni, J. *L'esperimento. Inchiesta sul Movimento 5 stelle*. Roma-Bari: Laterza, 2018.
Ilyenkov, E. Dialectics of the ideal. *Historical Materialism*, 20(2), 2012, 149–93.
James, W. *Pragmatism*. Indianapolis: Hackett Publishing Company, 1981.
Jullien, F. *The propensity of things. Towards a history of efficacy in China*, translated by J. Lloyd. New York: Zone Books, 1995.
Jullien, F. *Un sage est sans idée, ou l'autre de la philosophie*. Paris: Seuil, 1998.
Jullien, F. *Detour and access. Strategies of meaning in China and Greece*, translated by S. Hawks. New York: Zone Books, 2000.
Jullien, F. *Si parler va sans dire. Du logos et d'autres ressources*. Paris: Seuil, 2006.
Jullien, F. *De l'universel, de l'uniforme, du commun et du dialogue entre les cultures*. Paris: Fayard, 2008.
Jullien, F. *L'invention de l'ideal et le destin de l'Europe*. Paris: Seuil, 2009.
Jun, N.J. (ed.). *Brill's companion to anarchism and philosophy*. Leiden: Brill, 2017.
Jun, N.J., Wahl, S. (eds.). *New perspectives on anarchism*. Lanham: Rowman and Littlefield, 2010.
Keane, J. *The life and death of democracy*. London: Simon & Schuster, 2009.
Kant, I. *Critique of pure reason* (1781), ed. by P. Guyer and A.W. Wood. Cambridge/New York: Cambridge University Press, 2005.

Kant, I. *Critique of practical reason* (1788), ed. by M.J. Gregor, introduction by A. Wood. Cambridge: Cambridge University Press, 1996.

Kant, I. *Critique of the power of judgment* (1790), ed. by P. Guyer. Cambridge/New York: Cambridge University Press, 2000.

Kelly, G.A. *Idealism, politics and history. Sources of Hegelian thought*. Cambridge: Cambridge University Press, 1969.

Kelly, K. *What technology wants*. New York: Penguin, 2010.

Kitchin, R. *The data revolution. Big data, open data, data infrastructures and their consequences*. London: Sage, 2014.

Kroes, P. *Technical artefacts. Creations of mind and matter: a philosophy of engineering design*. Dordrecht: Springer, 2012.

Kuokkanen, M. (ed.). *Idealization VII. Structuralism, idealization and approximation*. Poznań studies in the philosophy of the sciences and the humanities, 42, 1994.

Kurzweil, R. *The singularity is near. When humans transcend biology*. New York: Viking, 2005.

Latour, B. *Science in action. How to follow scientists and engineers through society*. Milton Keynes: Open University Press, 1987.

Latour, B. *We have never been modern*. Cambridge: Harvard University Press, 1993.

Leonard, T. *Hewlett-Packard webcams "racist"*, 2009. https://www.telegraph.co.uk/technology/news/6874153/Hewlett-Packard-webcams-racist.html.

Levy, D. The ethical treatment of artificially conscious robots. *International Journal of Social Robotics*, 3(1), 2009, 209–16.

Like the flu? Trump's coronavirus messaging confuses public, pandemic researchers say. "The Guardian," March 13, 2020. https://www.theguardian.com/science/2020/jan/27/china-coronavirus-who-to-hold-special-meeting-in-beijing-as-death-toll-jumps.

Lloyd, G.E.R., Sivin, N. *The way and the word. Science and medicine in early China and Greece*. New Haven/London: Yale University Press, 2002.

Locke, J. *An essay concerning human understanding* (1690), ed. by P. Phemister. New York: Oxford University Press, 2008.

Lohr, S. *Facial recognition is accurate, if you're a white guy*. 2018. https://www.nytimes.com/2018/02/09/technology/facial-recognition-race-artificial-intelligence.html.

Longo, G.O. *Homo technologicus*. Roma: Meltemi, 2001.

Longxi, Z. *The tao and the logos. Literary hermeneutics, East and West*. Durham: Duke University Press, 1992.

Lovejoy, A.O. *The revolt against dualism. An inquiry concerning the existence of ideas*. La Salle: The Open Court Publishing Company, 1955.

Mace, S.E. (ed.). *Observation medicine. Principles and protocols*. Cambridge: Cambridge University Press, 2017.

Marcuse, H. *One-dimensional man. Studies in the ideology of advanced industrial society*, introduction by D. Kellner. London: Routledge, 1991.

Marshall, P.H. *Demanding the impossible. A history of anarchism*. Oakland: PM Press, 2010.

Marx, K., Engels, F. *Manifesto of the communist party*. New York: International Publishers, 1991.

May, T. *The political philosophy of poststructuralist anarchism*. University Park: Pennsylvania State University Press, 1994.

McCormick, P. (ed.). *Starmaking. Realism, anti-realism, and irrealism*. Cambridge: MIT Press, 1996.

McEwan, I. *Machines like me and people like you*. London: Jonathan Cape, 2019.

McLaughlin, P. *Anarchism and authority. A philosophical introduction to classical anarchism*. New York: Routledge, 2007.

McNally, P., Inayatullah, S. The rights of robots. Technology, law and culture in the 21st century. *Futures*, 20, 1998, 119–36.

Mittelstadt, B.D., Allo, P., Taddeo, M., Wachter, S., Floridi, L. The ethics of algorithms. Mapping the debate. *Big Data and Society*, 3(2), 2016, 1–21.

Moore, J. What is computer ethics? *Metaphilosophy*, 16(4), 266–75.

Morton, T. (ed.) *A Routledge literary sourcebook on Mary Shelley's Frankenstein*. London/New York: Routledge, 2002.

Murray, J. *A classical manual, being a mythological, historical, and geographical commentary on Pope's Homer and Dryden's Virgil*. London: Murray, 1833.

Newman, S. *Postanarchism*. Cambridge: Polity Press, 2016.

Nietzsche, F.W. *The gay science*, translated by W. Kaufmann. New York: Vintage, 1974.

Noble, S.U. *Algorithms of oppression. How search engines reinforce racism*. New York: New York University Press, 2018.

Nowak, L. Abstracts are not our constructs. The mental constructs are abstracts. In *Idealization I. general problems*, ed. by J. Brzeziński, F. Coniglione, T.A.F. Kuipers, L. Nowak. Poznań studies in the philosophy of the sciences and the humanities, 16, 1990, pp. 193–206.

Nowakowa, I., Nowak, L. (eds.). *Idealization X. The richness of idealization*. Poznań studies in the philosophy of the sciences and the humanities, 69, 2000.

Nozick, R. *Anarchy, state, utopia*. New York: Basic Books, 1974.

Nuzzo, A. *Ideal embodiment. Kant's theory of sensibility*. Indianapolis: Indiana University Press, 2008.

O'Neil, C. *Weapons of math destruction. How big data increases inequality and threatens democracy*. New York: Crown, 2016.

Ovid. *Metamorphoses*, translated by S. Lombardo, introduction by W.R. Johnson. Indianapolis/Cambridge: Hackett Publishing Company, 2010.

Panofsky, D., Panofsky, E. *Pandora's box. The changing aspects of a mythical symbol*. Princeton: Princeton University Press, 1978.

Pasquale, F. *The black box society. The secret algorithms behind money and information*. Cambridge: Harvard University Press, 2015.

Pasqualotto, G. *East & West. Identità e dialogo interculturale*. Venezia: Marsilio, 2003.

Pirandello, L. *One, no one and one hundred thousand*, ed. by W. Weaver. New York: Marsilio, 1992.

Plato. *Phaedrus*, ed. by R. Waterfield. Oxford/New York: Oxford University Press, 2009.

Plato. *Protagoras*, translated by B. Jowett, revised by M. Ostwald, ed. by G. Vlastos. Indianapolis/New York: Bobbs-Merrill, 1956.

Plato. Protagoras. In *Plato in twelve volumes*, vol. 3, translated by W.R.M. Lamb. Cambridge: Harvard University Press, 1967.

Plato. *Respublica*, ed. by P. Shorey. Cambridge: Harvard University Press, 1969.

Plato. *Symposium*, ed. by R. Waterfield. Oxford/New York: Oxford University Press, 2008.

Plato. Theaetetus. In *Plato in twelve volumes*, vol. 12, translated by H.N. Fowler. Cambridge: Harvard University Press, 1921.

Populism, politics and measles. "The New York Times," May 2, 2017. https://www.nytimes.com/2017/05/02/opinion/vaccination-populism-politics-and-measles.html.

Purves, D., Jenkins, R., Strawser, B.J. Autonomous machines, moral judgments, and acting for the right reasons. *Ethical Theory and Moral Practice*, 18(4), 2015, 851–72.

Rees, M. *On the future. Prospects for humanity*. Princeton/Oxford: Princeton University Press, 2018.

Reichenbach, H. *The rise of scientific philosophy*. Berkeley/Los Angeles: University of California Press, 1959.

Rescher, N. *Ethical idealism. An inquiry into the nature and function of ideals*. Berkeley/Los Angeles/New York: University of California Press, 1987.

Richter, H. *Dada. Art and anti-art*. London: Thames & Hudson, 1965.

Rockmore, T., Singer, B.J. (eds.). *Antifoundationalism old and new*. Philadelphia: Temple University Press, 1992.

Rorty, R. *Consequences of pragmatism. Essays, 1972–1980*. Minneapolis: University of Minnesota Press, 1982.

Rorty, R. *Contingency, irony, and solidarity*. Cambridge: Cambridge University Press, 1989.

Rosanvallon, P. *Counter-democracy. Politics in an age of distrust*. Cambridge: Cambridge University Press, 2008.

Rose, A. *Are face-detection cameras racist?* 2010. http://content.time.com/time/business/article/0,8599,1954643,00.html.

Rosenberg, N. *Exploring the black box. Technology, economics and history*. Cambridge: Cambridge University Press, 1994.

Ross, A. *The industries of the future*. New York: Simon & Schuster, 2016.

Ross, D. *Plato's theory of ideas*. Oxford: Clarendon, 1951.

Russell, B.A.W. *History of modern philosophy and its connection with political and social circumstances from the earliest time to the present day*. London: George Allen & Unwin LTD, 1946.

Russell, B.A.W., Whitehead, A.N. *Principia mathematica*. Cambridge: Cambridge University Press, 1963.

Santoro, G. *Un Grillo qualunque. Il Movimento a 5 stelle e il populismo digitale nella crisi dei partiti italiani*. Roma: Castelvecchi, 2012.

Sartwell, C. *Against the state. An introduction to anarchist political theory*. Albany: SUNY Press, 2008.

Schenk, H.G. *The mind of the European romantics*. London: Constable, 1966.

Schiller, J.C.F. *The ideal and the actual life*. In Id., *The poems and ballads of Schiller*, translated by Sir E. Bulwer Lytton. Leipzig: Tauchnitz, 1844.

Schiller, J.C.F. Letters upon the aesthetic education of man. In Id., *Literary and philosophical essays. French, German and Italian*. New York: Collier, The Harvard Classics, 1910.

Shakespeare, W. *Hamlet*. Oxford/New York: Oxford University Press, 1998.

Sharkey, N.E. The ethical frontiers of robotics. *Science*, 322, 2008, 1800–1.

Shelley, M. *Frankenstein. Or, the modern Prometheus*. London: Penguin Books, 1994.

Shelley, P.B. *Prometheus unbound. The text and the drafts: towards a modern definitive edition*, ed. by L.J. Zillman. New Haven/London: Yale University Press, 1968.

Shun, K., Wong, D.B. (eds.). *Confucian ethics. A comparative study of self, autonomy, and community*. New York: Cambridge University Press, 2004.

Smith, D. *Exceptional technologies. A continental philosophy of technology*. London: Bloomsbury, 2018.

Smith, W. (ed.). *Dictionary of Greek and Roman biography and mythology*. Boston: Little, Brown & Co., 1867.

Sophocles. *Oedipus the king*, translation by F. Storr. Cambridge: Harvard University Press, 1981.

Tatarkiewicz, W. *A history of six ideas. An essay in aesthetics*. The Hague: Nijhoff, 1980.

Tegmark, M. *Life 3.0. Being human in the age of artificial intelligence*. London: Allen Lane, 2017.

Terrone, E. *Filosofia dell'ingegneria*. Bologna: Il mulino, 2019.

The coronavirus crisis has brought the EU's failings into sharp relief. "The Guardian," March 29, 2020. https://www.theguardian.com/business/2020/mar/29/the-coronavirus-crisis-has-brought-the-eus-failings-into-sharp-relief.

The holy bible, Wheaton, Crossway, 2001.

Torrance, S. Ethics and consciousness in artificial agents. *AI & Society. Journal of Knowledge, Culture and Communication*, 22, 2008, 495–521.

Torres, P. *Morality, foresight, and human flourishing. An introduction to existential risks*. Durham: Pitchstone Publishing, 2017.

Trabattoni, F. *Platone*. Roma: Carocci, 1998.

Tronconi, F. (ed.). *Beppe Grillo's Five Star Movement. Organisation, communication and ideology*. Farnham: Ashgate, 2015.

Turing, A.M. Computing machinery and intelligence. *Mind*, 49, 1950, 433–60.

Tzara, T. Dada manifesto. In M.A. Caws (ed.), *Manifesto. A century of isms*. Lincoln/London: University of Nebraska Press, 2001, pp. 297–304.

Vaidhyanathan, S. *The googlization of everything (and why we should worry)*. Berkeley/Los Angeles: University of California Press, 2011.

van de Poel, I., Goldberg, D.E. (eds.). *Philosophy and engineering. An emerging agenda*. Dordrecht: Springer, 2010.

van de Poel, I., Royakkers, L. *Ethics, technology, and engineering. An introduction*. Oxford/Malden: Wiley-Blackwell, 2011.

Varanini, F. *Macchine per pensare. L'informatica come prosecuzione della filosofia con altri mezzi*. Milano: Guerini, 2015.
Vermaas, P., et al. *Philosophy and design. From engineering to architecture*. Dordrecht: Springer, 2008.
Vernant, J.-P. *Myth and society in ancient Greece*, translated by J. Lloyd. Brighton: Harvester Press, 1980.
Wallach, W., Allen, C. *Moral machines. Teaching robots right from wrong*. Oxford: Oxford University Press, 2008.
Weber, M. *Economy and society. An outline of interpretative sociology*. Berkley/Los Angeles: University of California Press, 2013.
Wittgenstein, L. *Tractatus logico-philosophicus*, translated by D.F. Pears and B.F. McGuinness, introduction by B. Russell. London: Routledge & Kegan Paul, 1963.
Wittgenstein, L. *Philosophical investigations*, ed. by P.M.S. Hacker and J. Schulte. Oxford: Wiley-Blackwell, 2009.
Wolff, R.P. *In defence of anarchism*. New York: Harper and Row, 1970.
Zellini, P. *La dittatura del calcolo*. Milano: Adelphi, 2018.

Index

"100 percent democracy," 6

abilities, x, 30, 77–78, 80, 89, 91, 102–4
absolute liberty, 110–12
abstraction, 33, 34, 36, 49n1, 80, 89, 90
Aeschylus, ix, 1, 5–7, 10, 11, 19, 35, 55, 77, 99, 102, 103, 125
aesthetics, 36, 37, 86
Against Method. Outline of an Anarchistic Theory of Knowledge (Feyerabend), 42
algorithm and health, 51–60
algorithmic machine, x, xi, 75, 78, 88–91, 98–100
Amazon, 21
ambition, 15, 16, 44, 94
analytic philosophy, 87
anarchism, 18, 20, 25, 28, 38–42, 45, 48, 55, 57–60, 62, 65, 66, 75, 83, 106, 107, 129, 130; epistemological, ix, 43, 44, 46, 47, 52, 54, 67–69, 71, 110; ethical, 52, 69; past and present forms, 109–14; political, 126n1; radical, ix, vii, xi, 114–26; technological, 74, 114, 124, 126
anonymity, 23, 60, 62, 69
antiquity and modernity, 17–19
anti-vaccination platform, 67
Aristotle, 80, 81
art: avant-garde, 42; contemporary, 1–2, 23, 116; work of, 37
artificial intelligence, 23, 24
The Art of Rhetoric (Aristotle), 81
Arts and Humanities Citation Index of Web of Science, 112
Athena, 7, 12, 97–99
authoritarianism, 67
autocracy, 60, 61, 66, 83
automation, xiii, 112, 119, 120, 125, 126
automaton materiale, xii, xiii, 113–26
autonomy, xii, xiii, 59, 60, 83, 84, 106, 107, 108n5, 110–14, 117, 122, 123, 125

Berlin, I., 36
bioengineering technologies, 97, 100
bureaucracy, ix, xi, 56–58, 112

Casaleggio, G., 72n9
Cassirer, B., 84
censorship, 111
The circle (Eggers), 21–22
cogitare, 84
Consequences of Pragmatism. Essays, 1972-1980 (Rorty), 43
contemporary medicine, 53
contingency, ix, 42, 44, 60, 63, 71

Copernican revolution, 82, 83
coronavirus emergency, xiii, 68, 125, 129, 130
creation, ix, xii, 29, 116, 117
Crime and Punishment (Dostoyevsky), 122
Critique of Pure Reason (Kant), 31, 82, 113
Critique of the Power of Judgment (Kant), 32
culture: European, 41, 121; Far Eastern, 117; Greek, 14, 25, 27, 126; history, 79; Western, vii–x, 1, 27, 28, 30, 31, 33, 34, 36, 39, 40, 41, 44, 55, 71, 75, 79, 84, 86, 87, 105–7, 109, 115, 116, 118, 122–4

data processing, 93
data science technologies, 98, 100
decision, 51, 53, 56, 63, 70, 71, 77, 78, 88–90, 100, 105, 107, 114, 117; epistemological, x, 69, 75; ethical, x, 69, 75, 117, 119; unilateral, 65, 67
decision-making process, 23, 52, 62, 78, 89, 111, 112
Deleuze, G., 109
Demiurge, 29
Descartes, R., 83, 107
Dialogue Concerning the Two Chief World Systems Ptolemaic and Copernican (Galilei), 81
digital platform, 65
digital technology, 65, 68
direct democracy, 63, 65, 67
A Discourse on Method (Descartes), 83
Distributed Autonomous Organization (DAO), 62
divine, xii, 35, 116, 117, 121; elements, 7, 15; fire, 2–3; immanent, ix, 20, 21, 25, 40–48, 53, 55, 56, 69, 118, 122; intelligence, 81; status, 17; transcendent, 47, 122
Dostoyevsky, F. M., 122
dualism, 28, 30, 31, 34, 36, 38, 40
dystopian scenario, 57

economic competence, 63
Eggers, D., 21–24, 57, 67
Einstein's Theory of Relativity (Cassirer), 84
Enlightenment culture, 69
Epimetheus, 2, 9, 101
epistemology, 60; evolution, ix; externalization, 53, 54; mistakes, 15, 16; phase, 28, 33; process, 34; revolution, 36–40; structure, 27, 33, 36–38, 40, 41; Western, 44, 46, 79, 80, 116
essential philosophical meanings, 68–69
Ethereum, 61, 62
ethical consequences, 16, 53, 54
ethical externalization, 55
ethics, 32, 86, 112, 117
European Academies of Fine Arts, 111, 112
European Conservatives and Reformists (ECR), 64
European human being, xii, 117, 118
Europe for Freedom and Direct Democracy (EFDD), 64
evils, 9, 10, 97, 110, 113
extremism, 62, 63, 67, 129, 130

Farewell to Reason (Feyerabend), 42
Feyerabend, P. K., 42
Fichte, J. G., 37, 38
fire, 2–5, 7, 9, 14, 15, 17, 35, 97–99, 101
Five Star Movement, 63–68
Foucault, M., 109
Frankenstein. Or, The Modern Prometheus (Shelley), 13–17
free association of citizens, 66
freedom, 44, 55, 66, 113, 114
French Revolution, 69
future prediction, 11–12

Galilei, Galileo, 81, 83
globalization, 130
God, 47, 121, 122
Godwin, W., 110

Goodman, N., 114–16
Google, 21, 45–48, 119, 120
Greek philosophy, 27
Greenfield, A., 63
Grillo, B., 64, 67, 72n9

Hamlet (Shakespeare), 70, 73–75, 102
Hegel, G. W. F., 36–38
Hellman, G., 115
Hephaestus, 7, 12, 97–99
herd immunity, 68
Herzen, A., 36
Hesiod, 1–4, 9, 121, 130
Hewlett-Packard, 21
historical evolution, ix, 1, 2, 13, 21, 25, 27, 28, 81, 92, 97, 120
The Holy Bible, 121
Horizon 2020 program, 91–92
human, 6, 9, 100–102; brain, 92–93; and divine, viii, 3–5, 11–13, 15, 20, 21, 23, 39, 40, 44–45, 84; identity, 11, 78, 81–85, 103; intellect, 81, 82; life, 10; *vs.* living creatures, 2, 3, 105; mind, 23, 81, 89–94, 101, 102; and nature, 7; reason, 32, 34, 82; sacrifice, 36; and technology, 11–14, 16–19, 34, 35, 39, 44; Western, vii, x, xii, 31, 34, 35, 68–70, 93, 105, 106, 107, 117, 126
Human Brain Project (HBP), 92
The Human Condition (1933), 116
humanity, 25, 32
hypertrophization of *logos,* x–xi, 79–94
The Ideal and the Actual Life (Schiller), 37

idealism, 49n6
ideality, ix, 57, 71, 80, 87, 105, 106, 124; ancient invention of, 27–36; modern crisis of, 36–40
idealization, 33, 80
imagination, 33, 34, 97, 102
immanence, 47, 57, 62, 124
incommensurability, 58, 59
individual conscience, 110–12
individual lives, 118
individual responsibility, 63, 117, 118, 120, 123
intelligent mastery, 2, 3, 5
irrealism, 114, 115
Italian Academy of Fine Arts, 111
Italian Constitution, 64
Italian Ministry of Health, 68

James, W., 109

Kant, I., xii, 31–35, 38, 69, 82, 83, 106, 107, 112–14, 117, 126
knowledge, 7, 12, 15, 16, 22, 47, 79, 80, 81, 115

languages, 115, 116
Locke, J., 33, 86–87
logical thought, 85, 87

Machines like Me and People like You (McEwan), 23
Magritte, R., 116
Massachusetts Institute of Technology, 72n16
mathematical sciences, 81, 82
McEwan, I., 23–24, 91
measles, mumps and rubella (MMR), 68
mechanical engineering technologies, 98, 100
mental attitudes, 2, 3
mental strength, 75, 77
meta-externalization, x, xi, 73–78, 88, 89, 91, 93, 102, 103–6, 112, 118, 123
Metamorphoses (Ovid), 7–8
metaphysics, 86
modern scientific method, 81
money, 61–71, 94
moral duties, 122
moral law, 110
Moral Machine experiment, 72n16
Murray, J., 125

narcissism, 44, 58
New York Times, 67

Nicomachean Ethics (Aristotle), 80
Nietzsche, F. W., 122
nihilism, 122
Nikon, 21
nudes, 111

objective knowledge, 66, 83
Oedipus the King (Sophocles), 70
Olympians, 9, 101
omnipresence, omniscience, and omnipotence, vii, viii, xii, 12, 20–22, 45–47, 61, 114, 120
online streaming, 69
ontological differences, viii, 11, 18, 19
ontological dimension, ix, 18, 20, 38–40, 44, 45, 47, 48, 58, 124
ontological quality, 7, 8, 11, 28, 38, 39, 42, 44, 47
ontological status, x, xii, 28–30, 77, 94, 103, 104, 115, 116, 120–23
Ovid, 1, 7–8, 100, 101

Pandora, 9, 13, 25, 97
particularity, 28, 39, 42–44, 66, 71, 80, 126
personality, 37
philosophy: of history, 36, 37; political, 38
Pirandello, L., ix, 70–71, 78, 107, 114, 116, 117
Plato, 1–3, 7, 12, 28–33, 35, 79–80, 91, 103, 107, 117, 129, 130
political alliances, 64, 66, 69
political party, 64, 66
Populism, Politics, and Measles, 67
populist politicians, 68
post-truth, 68
power management, 56
predictive algorithm, 51–54, 56, 78, 90
private judgment, 110–12
Prometheus, 10, 100–102, 105, 121; ancient, 1–13, 15–17, 19–21, 25, 34, 39, 121, 123; creating humans, 8; deceit, 4, 9; figure of, ix, viii, 1–2, 14, 21, 25, 27, 28, 36, 81, 92, 97, 120; gifts, 5, 7; modern, 13–25, 44–46, 67, 121; myth of, 2–4, 9–13
Prometheus Bound (Aeschylus), 5, 6, 10, 11, 19, 35
Protagoras (Plato), 2, 3
pseudonymity, 62, 69
psychological strategy, 59, 60
public funding, 64, 66, 69
punishment, 4, 17, 18, 113, 114

radical externalization, x, 73–78
radicalization, viii, 40, 59, 83, 86, 87, 109, 110
rationality, xi, 56, 88, 89, 93, 94, 117
reality, ix, xi,16, 27, 28, 35–42, 44, 57, 71, 80, 85, 87, 88, 104, 105
Reichenbach, H., 73–75, 78, 88, 91, 94, 107, 117
relativism, 41, 42, 71
relativistic physics, 84
representative democracy, 63–65, 67
Republic (Plato), 28–29
romanticism, 37
Rorty, R., 43
Rousseau, 65
Russell, B. A. W., 86–88, 118

Schiller, J. C. F., 37, 38, 40
scientific revolution, 81–84
search engine, 119
SENS project, 6
Shakespeare, William, ix, 70, 73, 74, 78, 102, 107, 117
Shelley, M., 1, 13–15, 23, 36, 116
Shelley, P. B., 13, 19, 35, 36
Smith, W., 125
social media/networks, xiii, 58, 59, 68, 111, 112, 129
solipsism, 46, 58, 60, 61, 67, 69, 71
Some Lectures Concerning the Scholar's Vocation (Fichte), 37–38
Sophocles, 70, 78, 107, 117
structured language, 115
subjective knowledge, 66, 83
superior ontology, ix, 35, 39, 40, 44

technological era, vii, viii, x–xii, 1, 6, 10, 11, 13, 16–21, 25, 28, 39, 40, 42, 45, 55, 58, 71, 105–7, 116–18, 120, 122, 124
technology, 4, 7, 8, 11, 14–17, 24, 75–77, 92, 107, 125; block, 112; consequences of, 5–6; contemporary, 10, 40–48, 123; and humans, 11–14, 16–19, 34, 35, 39, 44; identities, 63; ontology, 9, 11; power, 104; Western notion of, 19, 20, 23; working method of, 29, 31, 33. *See also individual entries*
Theaetetus (Plato), 79
Theogony (Hesiod), 2, 4, 121
theory of relativity, 84
Titan, 9, 16, 18
"Tiziano e Caravaggio in Peterzano" (exhibition), 111
totalitarianism, 28, 37, 38, 41, 50n7, 57, 62, 67, 129, 130
Tractatus Logico-Philosophicus (Wittgenstein), 85, 86
transcendence, 20, 58, 124
truth, 34, 46, 47, 83, 84, 86, 102, 118
Tsunami Tour, 64
Turing, A. M., 24, 91
Tzara, T., 43, 113

universal ideal matrix, 29
universality, 28, 38, 39, 43, 66, 71, 80

vaccinations, 68

Western history, 69
Western philosophy, 74, 79, 86, 87
wisdom: Athena's, 2, 7, 98, 99; civic, 12, 17, 25, 129–30; political, 12, 13, 17
Wittgenstein, L., 85–87, 107
Works and Days (Hesiod), 9
World Health Organization, 68

YouTube, 59

Zeus, 3–4, 9–12, 17, 19, 20, 25, 97, 99–102, 105, 121

About the Author

Simona Chiodo is a full professor of philosophy at the Politecnico di Milano and a member of its Research Ethical Committee. She is the author of several monographs and peer-reviewed journal articles on epistemology (in particular, the relationship between reality and ideality), as well as on aesthetics, and coeditor of the recent volume *Italian Philosophy of Technology*.

www.ingramcontent.com/pod-product-compliance
Ingram Content Group UK Ltd.
Pitfield, Milton Keynes, MK11 3LW, UK
UKHW041843031026
15587UKWH00003B/983

9 781793 632944